Synthesis Lectures on Ocean Systems Engineering

Series Editor

Nikolas Xiros, University of New Orleans, New Orleans, LA, USA

The series publishes short books on state-of-the-art research and applications in related and interdependent areas of design, construction, maintenance and operation of marine vessels and structures as well as ocean and oceanic engineering.

Alexander Arnfinn Olsen · Fidaa Karkori

Development of Procedures and Technical Manuals for the Marine and Offshore Industries

Alexander Arnfinn Olsen ⓘ
Southampton, UK

Fidaa Karkori
Southampton, UK

ISSN 2692-4420　　　　　　ISSN 2692-4471　(electronic)
Synthesis Lectures on Ocean Systems Engineering
ISBN 978-3-031-74862-2　　　ISBN 978-3-031-74863-9　(eBook)
https://doi.org/10.1007/978-3-031-74863-9

© The Editor(s) (if applicable) and The Author(s), under exclusive license to Springer
Nature Switzerland AG 2025

This work is subject to copyright. All rights are solely and exclusively licensed by the Publisher, whether the whole or part of the material is concerned, specifically the rights of translation, reprinting, reuse of illustrations, recitation, broadcasting, reproduction on microfilms or in any other physical way, and transmission or information storage and retrieval, electronic adaptation, computer software, or by similar or dissimilar methodology now known or hereafter developed.
The use of general descriptive names, registered names, trademarks, service marks, etc. in this publication does not imply, even in the absence of a specific statement, that such names are exempt from the relevant protective laws and regulations and therefore free for general use.
The publisher, the authors and the editors are safe to assume that the advice and information in this book are believed to be true and accurate at the date of publication. Neither the publisher nor the authors or the editors give a warranty, expressed or implied, with respect to the material contained herein or for any errors or omissions that may have been made. The publisher remains neutral with regard to jurisdictional claims in published maps and institutional affiliations.

This Springer imprint is published by the registered company Springer Nature Switzerland AG
The registered company address is: Gewerbestrasse 11, 6330 Cham, Switzerland

If disposing of this product, please recycle the paper.

The future of the safety movement is not so much dependent upon the invention of safety devices as on the improvement of methods of educating people to the ideal of caution and safety.

Walter Dill Scott

Preface

Safety and environmental concerns demand an investment in time and energy to develop and implement accurate, effective instructional materials. Consistency is the key to efficiency, quality control, and safe operations. This consistency can come from the use of well-written administrative, regulatory, operating, emergency, maintenance, and routine duty procedures and technical manuals. The purpose of this guidance is to aid in aligning instructional material to safety and environmental considerations; revising existing documents for conformance to current practices, policies, and regulatory requirements; and implementing the new and/or updated documents into a safety program to enhance productivity and reduce accidents. The guidance presented is recommendatory. Although it is widely recognised the maritime industry encourages owners, operators, and manufacturers to adopt these principles wherever feasible and applicable, compliance is not required. The intent is for technical documentation writers to tailor the information presented in this guidance to the needs and preferences of their particular organisation.

Southampton, UK Alexander Arnfinn Olsen
2024 Fidaa Karkori

Acknowledgements It is with the deepest gratitude that we thank the team at Babcock Marine and Technology for sharing their knowledge and insights during the development of this guide. I would also like to extend my thanks to the editorial team at Springer for their assistance in bringing this guide together.

To you all, our sincerest thanks and gratitude.

Contents

1	**Introduction**		1
	1.1	General	1
		1.1.1 Application	1
		1.1.2 Definitions for Instructional Materials	2
	1.2	Definitions of a Procedure	3
		1.2.1 Compliance with Standards	3
	1.3	Definitions of a Technical Manual	4
	1.4	Importance of Instructional Materials	4
	1.5	Influencing Compliance and Performance	5
	1.6	Structure of this Guidance	5
2	**The Role of Instructional Materials**		7
	2.1	General	7
		2.1.1 Definitions	7
		2.1.2 Limits of Written Materials Relating to Maritime Environments	7
	2.2	Purpose of Instructional Materials	8
	2.3	Uses of Instructional Materials	9
	2.4	Concerns to Address in Instructional Materials	9
		2.4.1 High-Level Concerns	9
		2.4.2 Safe Operating Limits	10
	2.5	Types of Instructional Materials	10
	2.6	Management Oversight of Instructional Materials	11
3	**Writing Instructional Materials**		13
	3.1	General	13
		3.1.1 Definitions	14
	3.2	Initiations of New Instructional Materials	15
	3.3	Goal Definition	15

	3.4	Management Approval to Proceed	15
	3.5	Information Gathering	16
		3.5.1 Subject Matter Experts (SME)	16
		3.5.2 Source/Supporting Documentation	16
		3.5.3 Writing to a Specific Audience	17
	3.6	Effective Presentation of Material	17
	3.7	Enhancing Usability	18
		3.7.1 User Participation	18
		3.7.2 Instructional Material Effectiveness	18
		3.7.3 Instructional Material Maintenance	18
	3.8	Instructional Material Writing Elements	19
		3.8.1 Instruction Number	19
		3.8.2 Clarity	19
		3.8.3 Brevity	20
		3.8.4 Use of Imperative (Command) Sentence Structure	20
		3.8.5 User Friendliness/Usability	20
		3.8.6 Technical Accuracy	20
	3.9	Style Format Guidelines	21
		3.9.1 Font	21
		3.9.2 Sentence Structure	21
		3.9.3 Vocabulary	24
		3.9.4 Level of Detail	26
	3.10	Document Organisation	26
		3.10.1 Front Matter	27
		3.10.2 Document Sections	31
		3.10.3 Ending the Procedure	33
		3.10.4 Appended Information	33
4	**Graphics in Instructional Materials**		35
	4.1	General	35
		4.1.1 Definitions	35
		4.1.2 Graphics	36
		4.1.3 Sources of Graphics	36
		4.1.4 Types of Graphics	37
		4.1.5 Guidelines for Using Graphics in Instructional Materials	38
		4.1.6 Incorporating Electronic Media into Instructional Materials	40
5	**Verifying, Validating, Approving, Certifying, and Implementing Instructional Materials**		41
	5.1	General	41
		5.1.1 Definitions	41

		5.1.2	Final Phase of Instructional Document Development Process	42
	5.2	Verifying and Validating Instructional Materials		42
		5.2.1	Verification	42
		5.2.2	Validation	42
	5.3	Resolving and Incorporating Final Comments		43
	5.4	Proofreading and Checking for Format and Style		43
		5.4.1	Proofreading	44
		5.4.2	Instructional Document Format and Style	44
	5.5	Approving the New/Revised Instructional Materials		44
	5.6	Certifying the New Instructional Materials		44
	5.7	Implementing New/Revised Instructional Materials		44
		5.7.1	Planning for Implementation	45
		5.7.2	Distribution	45
6	**Managing Instructional Materials**			47
	6.1	General		47
		6.1.1	Document Control	47
		6.1.2	Access	48
		6.1.3	Training	48
	6.2	Maintenances of Instructional Materials		48
		6.2.1	Updates	48
		6.2.2	Periodic Reviews	49

Appendix A: Checklist for the Preparation of Instructional Materials 51

Appendix B: Example Constrained Word List and Constrained Language List 53

Appendix C: Analysis Technique 75

Appendix D: Examples of Instructional Materials 81

Bibliography 85

Abbreviations and Acronyms

COLREGS	International Regulations for Preventing Collisions at Sea
HAZID	Hazard Identification
HAZOP	Hazard and Operability Analysis
IMO	International Maritime Organisation
ISO	International Organisation for Standardisation
JHA	Job Hazards Analysis
MSC	Maritime Safety Committee
MSDS	Material Safety Data Sheet (MSDS)
PPE	Personal Protective Equipment
SME	Subject Matter Expert

Phonetic Alphabet and Numbers

A	Alpha	Al Fah
B	Bravo	Brah Voh
C	Charlie	Char Lee
D	Delta	Dell Tah
E	Echo	Eck Oh
F	Foxtrot	Foks Trot
G	Golf	Golf
H	Hotel	Ho Tell
I	India	In Dee Ah
J	Juliet	Jew Lee Ett
K	Kilo	Key Loh
L	Lima	Lee Mah
M	Mike	Mike
N	November	No Vem Ber
O	Oscar	Oss Cah
P	Papa	Pah Pah
Q	Quebec	Keh Beck
R	Romeo	Row Me Oh
S	Sierra	See Air Rah
T	Tango	Tang Go
U	Uniform	You Nee Form
V	Victor	Vik Tah
W	Whisky	Wiss Key
X	Xray	Ecks Ray

Y	Yankee	Yang Key
Z	Zulu	Zoo Loo

1	One	Wun
2	Two	Too
3	Three	Tree
4	Four	Fow Er
5	Five	Fife
6	Six	Six
7	Seven	Sev En
8	Eight	Ait
9	Nine	Nin Er
0	Zero	Ze Ro

List of Figures

Fig. 3.1	Sample template	28
Fig. 4.1	Three (3) dimensional diagram	38
Fig. 4.2	Flowchart showing a process	39
Fig. C.1	Example form for task analysis for instructional materials	79

Introduction 1

1.1 General

Humans have a limited capacity to store information in short-term and working memory. Research suggests that humans can process in short-term or working memory about five to nine pieces of information at one time. Working memory is that memory involved in directed conscious attention. For example, start-up operating instructions for the manual operation of a potable water system requires the operator to remember pumps, chemical levels, etc. For this reason, the maritime and other industries should not rely exclusively on an individual's memory to perform work-related tasks, especially those that are complex, hazardous, and have the potential to impact personnel safety, equipment, or the environment. Appropriately written and implemented instructional materials (procedures and technical manuals) can reduce the cognitive effort of personnel, especially the memory element of performing a task, thus aiding in reducing human errors.

1.1.1 Application

The principles and guidelines in this guidance applies to writers of maritime procedures and technical manuals who wish to further understand the document development process, the intent of, and the benefits derived from appropriate procedure and technical manual design, development, implementation, and life-cycle tracking. For the purpose of brevity, the terms "procedures" and "technical manuals" are not explicitly mentioned throughout this document and are instead referred to as "instructional materials/documents." These umbrella terms include procedures, technical manuals, checklists, and/or any other instructional-type of material used to perform a specific task(s). This guidance can be considered applicable to all instructional documents in both paper and electronic format.

This guidance is intended to serve as a reference for maritime organisations when planning, developing, implementing, revising, and maintaining instructional materials. This guidance describes general concepts, provide design considerations, and offer content suggestions for instructional materials. The goal of this guidance is for the writer to tailor the instructional material to their own unique needs and preferences, so long as the concepts of clarity, accuracy, and completeness are maintained while also promoting safety and efficiency.

1.1.2 Definitions for Instructional Materials

Administrative Procedure: Developed to meet an organisation's goals, including planning, organising, staffing, directing and motivating personnel, controlling quality, and budgeting.

Checklist: A list of actions or events to be noted, checked, or remembered, and when checked, verifies that the actions have been performed.

Emergency Procedure: A list of steps that describe required actions or tasks to be taken in the event of a failure of the equipment or system.

Human Error: Performance of humans that deviates from the desired performance. This definition is not a failure to perform as directed, but failure to perform as desired.

Instructional Material/Document: An umbrella term used to describe well-defined procedural documents, technical manuals, and/or any other written material that is intended for training or learning of tasks, processes, procedures, and systems. These terms also include instructional materials in any digital or electronic format.

Job Aid: Includes any device (whether in written, mechanical, electronic, or other form) that can be used by a person to facilitate the performance of a job or task. Job aids are often printed or visual summaries of key points or steps essential to the performance of a task.

Maintenance Procedure: A list of steps that describes how to perform a task required to maintain or repair an equipment item. It is also a documented standard to which the job or task should be performed.

Management of Change: A systematic program to assure that changes to a process are appropriately reviewed, and any hazards introduced by the change are identified, analysed, and controlled prior to resuming operation.

Near Miss/Close Call: An event, or chain of events, that under slightly different circumstances could have resulted in an accident, injury, damage, or loss of personnel, equipment or the vessel.

Operating Procedure: Developed to include regularly recurring work processes.

Procedure: A fixed, step-by-step sequence of activities or course of action that have definite start and end points, and that must be followed in a specific order to perform a task correctly and safely.

Routine Duty Procedure: Developed for activities such as regular checks of gauges for indication of operating temperatures or pressures, checks for running (on/off) condition of equipment, and checks for unusual noises or vibrations.

Safe Operating Procedure: Those that are directed specifically towards the safe use of equipment or processes.

Technical Manual: A well-defined document containing procedures on operation, handling, maintenance, and repair.

Short-Term Memory: The mechanism for not only holding a limited amount of information in our consciousness in an active state but also readily available for a short period of time.

Technical Manual: A well-defined set of instructions that have a specific goal to be accomplished, such as installation, operation, usage, maintenance, and training for the effective use of equipment, machinery, or deployment of processes and systems.

Working Memory: The execution and attention aspect of short-term memory involved in the integration, processing, disposal, and retrieval of information.

1.2 Definitions of a Procedure

Procedures are a specified series of actions or operations leading to an outcome within the operation of a process or system. Procedures are detailed lists of steps describing how to perform various tasks, and as a result, they represent the quality to which the tasks should be performed. Procedures explain what is expected and required of a person when they perform a specific task. Procedures may range from detailed guidance to step-by-step instructions, to job aids, or short checklists. In order to determine the level of detail needed in a procedure, there are several factors to consider, namely the significance of error, the complexity of the task, how often the task is performed, as well as the knowledge, skills, and abilities of the user. Job aids can reduce the amount of decision-making required and decrease the need to memorise key points of the task. Job aids include any device that can be used by a person to facilitate the performance of a job or task. Job aids are often printed or visual summaries of key points or steps essential to the performance of a task. They assist by directing, guiding, and informing people. Some formats for job aids can be in the form of checklists, worksheets, decision tables, algorithms, etc. (refer to Appendix D, "Examples of Instructional Materials").

1.2.1 Compliance with Standards

It is important to note that some documented procedures must be prepared to comply with particular standards published by the International Organisation for Standardisation (ISO). For example, ISO 9001, this document specifies requirements for maintaining effective quality assurance systems for manufacturing and service industries. Additionally, the International Maritime Organisation (IMO) Maritime Safety Committee and other regulatory bodies can make amendments to older editions of specific procedural documents that are already in place. In a June 2013 meeting, there were amendments to the 1979, 1989, and 2009 editions of the MODU Code, which now require that specific procedures be developed for entry into enclosed spaces. This example reflects the importance, for the writers of instructional materials, to continuously monitor changes or amendments in industry standards.

1.3 Definitions of a Technical Manual

A technical manual is a well-defined document that explains the means for operating, maintaining, supporting or installing a machine, process, system, or piece of equipment. As a result, they represent the quality to which these various tasks should be performed and explain what is expected and required of a person when they perform a specific task. Compared to operating or maintenance procedures, technical manuals guide integration and installation of equipment into the processes of a larger system. They also guide detailed, complex, and infrequent maintenance activities such as extensive overhaul or update/replacement of subsystems/components of the larger device. Technical manuals also provide much of the technical information required to be incorporated into a company's operating, logistics, maintenance and administrative procedures. These technical concerns include the following types of activities: operating tolerances, specifications for consumables, inspection and maintenance requirements, and training requirements. They are much more detailed than procedures. Thus, in order to determine the level of detail needed in a manual, there are several factors to consider, namely the significance of error, the complexity of the task, how often the task is performed, as well as the knowledge, skills, and abilities of the user.

1.4 Importance of Instructional Materials

Instructional materials represent the link among administration personnel, management, manufacturers, supervisors, and the personnel who perform the various tasks described in these documents. When personnel perform their work in accordance with instructional materials, they implement the plans, agreements, and policies incorporated in the documents and required by management and regulators.

1.5 Influencing Compliance and Performance

Human performance (e.g., safe, efficient, and reliable task performance) can be monitored and influenced. An inadequate level of human performance can adversely impact operations and safety. The primary objective for influencing compliance with instructional materials and performance is the reduction of errors. Properly developed materials can influence human performance by reducing variation in work performance and facilitating compliance with the requirements of the documents. This is accomplished by documenting standard work processes in instructional materials. Human error can lead to near misses or minor losses, escalating up to major accidents. Inappropriate and undesirable decisions or behaviours can increase the likelihood of a loss. When developing instructional materials, it is important to focus on reducing the number of decisions that people have to make, as well as reducing risk and memory recall requirements. Instructional materials can reliably carry information concerning decision rules (e.g., "Vent steam when temperature is reduced to 126 °C". Risk perceptions can be mitigated via the use of danger, warning, and caution labelling and signage. Instructional documents can also aid in memory requirements by listing steps and their particulars, directing the when, where, how, why, and what of instructional steps, and indicate any factors related to task timing.

1.6 Structure of this Guidance

This guidance focuses on the process for developing and maintaining new and existing instructional materials. The structure is as follows:

- Chapter 2—The Role of Instructional Materials
- Chapter 3—Writing Instructional Materials
- Chapter 4—Graphics in Instructional Materials
- Chapter 5—Managing Instructional Materials
- Chapter 6—Verifying, Validating, Approving, Certifying, and Implementing Instructional Materials.

Appendices:

- Appendix A—Checklist for the Preparation of Instructional Materials
- Appendix B—Example Constrained Word List and Constrained Language List
- Appendix C—Analysis Techniques; and
- Appendix D—Examples of Instructional Materials.

The Role of Instructional Materials

2.1 General

The development and proper use of instructional materials are integral parts of a successful quality system because they provide people with the information needed to perform tasks appropriately, efficiently, and safely.

2.1.1 Definitions

Deviation: Any action or inaction that is outside the established safe tolerances, which can include exceeding a safe upper or lower limit, skipping a safety-critical step in a procedure, or performing a step or sequence of steps out of order or in an unsafe manner.

Operating Limits: The boundaries or tolerances in which a system is at safe operating conditions.

2.1.2 Limits of Written Materials Relating to Maritime Environments

It is not possible to create a set of instructions that perfectly address every situation. There are numerous operational and environmental contexts whereby an onboard procedure could be implemented as specified, implemented in a different way, or not implemented at all due to safety or environmental concerns. The standards and regulations of the maritime industry (shipping and offshore) can become overly complex, and adhering to all the various navigation and conning rules may become difficult to satisfy. As an example, consider this statement from the International Regulations for Preventing Collisions at Sea (COLREGS) by Rule 2, RESPONSIBILITY:

(a) Nothing in these Rules shall exonerate any vessel, or the owner, Master, or crew thereof, from the consequences of any neglect to comply with these Rules or of the neglect of any precaution which may be required by the ordinary practice of seamen, or by the special circumstances of the case.

(b) In construing and complying with these Rules, due regard shall be had to all dangers of navigation and collision and to any special circumstances, including the limitations of the vessels involved, which may make a departure from these Rules necessary to avoid immediate danger.

The above can also be applied to instructions and guidance expressed in instructional materials, where the ultimate goal is safe and efficient operations. "Directed performance" demands rigid compliance with the procedures. In contrast, "desired performance" should produce safe and efficient job performance. Within specific contexts, "directed performance" could be deemed unsafe by personnel and therefore application of "desired performance" would allow personnel to adapt to the situation and perform the task in a safe manner according to the specific situation.

2.2 Purpose of Instructional Materials

The purpose of instructional materials is to provide personnel with the information necessary to perform a task safely, efficiently, and uniformly, so that each person performs the same task the same way with the same result. The narrower function of individual, task-specific instructions is discussed in Chap. 3, "Writing Instructional Documents". Where applicable, instructional materials should provide direction on how to safely start, operate, and shut down equipment, while at the same time clearly specify when an emergency shutdown must be executed. Well-designed and well-written documents incorporate all safety issues related to a particular task, and therefore, maximise safe performance each and every time the task is performed due to the uniformity element built into the documents. This uniformity element includes the use of Dangers, Warnings, and Cautions in procedures to keep personnel aware of the safety issues related to the task being performed. For further details on where to place them in an instructional document, refer to Chap. 3, Table 3.1, "Usage of Signal Words".

2.3 Uses of Instructional Materials

When operations are recognised as being safety critical, involve lengthy task sequences beyond a reliable memory limit, are overly complex, or involve the presentation of complex information, there is a need for an instructional document. These tasks can involve normal operations, non-routine or infrequent tasks, low- to high-risk tasks, and situations where a task requires multiple people. When multiple people are required, the document should define their roles and responsibilities. Tasks described in instructional materials should keep in mind the knowledge, skills, and abilities of the personnel. When a qualified person is required to successfully perform a task consistently, especially one that challenges memory, an instructional document is needed. Tasks that are performed solely from memory can lead to unpredictable and unsafe outcomes. Instructional materials are also needed to:

- Satisfy regulatory requirements
- Satisfy ISO standard requirements if certified by ISO
- Respond to specific safety incidents; and
- Respond to specific process loss events.

2.4 Concerns to Address in Instructional Materials

The concerns discussed in the following Paragraphs should be addressed in instructional documents, depending on the specific task, the workplace environment, and the associated risk involved in task performance. Well-developed materials describe in sufficient detail the hazards, the tools required, the necessary protective equipment, and the specific control devices so that people are aware of potential hazards. In doing so, they can verify that controls are operational and confirm that the equipment responds as expected. If the system does not respond as expected, well-developed instructional materials include troubleshooting information and necessary actions when and where required.

2.4.1 High-Level Concerns

The list below displays some common high-level concerns that should be addressed in instructional materials:

- What are the specific tasks required of personnel to operate and maintain their equipment?
- Who are the required operating personnel and who are the required maintenance personnel?

- What will be the operating environment?
- What types of human error can be anticipated and what are the possible consequences of those errors?
- What are the safety concerns?

2.4.2 Safe Operating Limits

Where applicable, instructional materials should describe safe operating limits that are set for critical process parameters such as pressure level, temperature, flow, electrical power, etc., and conditions based on a combination of equipment design limits and the dynamics of the system. Safe operating limits are usually specified when the system response may be so severe that continued operation is inadvisable due to associated risks. Instructional documents related to these types of system responses should include clear, accurate instructions to respond to the situation. For each safe operating limit, the potential consequence of exceeding the limit should be included in the document, along with the steps to avoid deviation or return to a safe condition if an excursion outside of the safe operating limit occurs.

2.5 Types of Instructional Materials

There are many types of instructional materials used in the maritime industry which include many procedural and technical documents. Day-to-day operating categories for instructional materials include:

- Administrative (e.g., route or passage planning)
- Operating/Bridge, Engineering (e.g., watch turn-over, pilot briefings, engine start-up/shut-down)
- Emergency (e.g., emergency escape)
- Routine Duty (e.g., temperature or pressures checks, walkdowns/walkthroughs); and
- Alarm (e.g., man overboard).

Technical manuals can become permanent procedures or protocols of a company. The procedures, if developed further can become technical manuals. These documents guide the following:

- Integration and installation within the overall larger system
- Technical and operating limits
- Maintenance, test, and calibration schedules and requirements
- Major to minor overhauls

- Maintenance (e.g., routine maintenance tasks or repair maintenance tasks)
- Temporal tasks (e.g., conditional or infrequent operations); and
- Decommissioning.

2.6 Management Oversight of Instructional Materials

Management should develop a policy or formal program that describes the process for developing, updating, and maintaining instructional materials. This policy should define the range of activities to be considered for the documents, and the general roles and responsibilities of persons concerned with the development and maintenance of these documents. Management should be accountable for establishing a culture in which personnel willingly conform to the instructional materials. Management should also verify compliance with these documents by regularly observing their performance in the field (refer to Chap. 6, "Managing Instructional Documents").

Writing Instructional Materials 3

3.1 General

There are many steps in the development of instructional documents. Appendix A, "Checklist for the Preparation of Instructional Materials", provides an example checklist to follow when developing documents. The checklist outlines the development process from pre-development activities to managing documents in use. Prior to the development of an instructional document, several tasks should be performed so that the need for a document can be assessed, and the content identified and addressed. Some considerations that precede document development include the following:

- Understanding when and where instructional materials are required
- Understanding goals and general concepts of a task(s)
- Fully understanding the process and its interrelationship with other processes (process mapping)
- Gathering relevant information and identifying which documents are needed
- Assessing current company and industry resources for best practices
- Assessing current industry standards, including regulatory requirements
- Determining the need for materials through a Needs Assessment
- Performing a Risk Assessment, Task Analysis, or Gap Analysis; and
- Determining design and layout of the document for ease of use and distribution (i.e., paper copies or interactive electronic versions)

For information regarding optional pre-document analysis techniques, including how to conduct a needs assessment or task analysis, refer to Appendix C, "Analysis Techniques".

3.1.1 Definitions

Action Word: A powerful verb that describes the effort used to complete a task.

Active Voice: The subject of the sentence performs the action of the verb (e.g., Set up the test box on the power panel board).

Caution: In the context of instructional materials, a statement that indicates a potentially hazardous situation that, if not avoided, may result in a minor or major injury. Cautions may be used to signify situations where property damages or a minor pollution problem are possible.

Clause: A group of words containing a subject and a verb usually forming part of a compound or complex sentence.

Constrained Word List: A list of preferred words and terms shown adjacent to corresponding non-preferred words and terms (refer to Appendix B).

Danger: A statement describing an imminently hazardous situation that, if not avoided, will result in death or serious injury. This signal word is to be limited to the most extreme situations.

Hazard and Operability Analysis (HAZOP) : An analysis technique that uses special guidewords to prompt an experienced group of people to identify potential hazards or operability concerns relating to the operation of equipment or systems.

Maintenance: All activities necessary to keep equipment up to, or restored to, a specified level of performance.

Note: *Information that helps to explain a process but may not be needed for task performance; background information that may have a role in supporting decision-making.*

Personal Protective Equipment (PPE): Includes all clothing and other worn work accessories designed to create a barrier against workplace hazards. Examples include safety goggles, hard hats, hearing protectors, gloves, respirators, aprons, and work boots.

Step: In the context of an instructional document, a command statement consisting of information that requires an action from a person. The command statement includes an action word (verb), the object of the action, and often the location where the action is performed.

Style: The manner in which the author choses to write to his or her audience; this can make one document look different from another.

Style Guide: A document that describes and illustrates a prevailing, accepted, or authorised style.

Task: A necessary step which enhances or provides benefit to a process in some way; a job, step, or function within a process which leads to a desired output or result.

Usability: Convenience and practicability for use; ease with which information can be found and understood in a document.

User-Friendly: A user-focused information presentation style that makes complex information easy to access, read, use, and understand.

Validation: The assurance that a product, service, or system meets a set of design specifications based on the needs of the customer or client.

Verification: The evaluation of whether or not a product, service, or system complies with a regulation, requirement, specification, or imposed condition. This is often an external process.

Warning: A statement describing an abnormal situation where there is a possibility of injury to personnel and/or a major impact to the environment if the warning is not heeded.

3.2 Initiations of New Instructional Materials

The first step in initiating a new instructional document is the recognition of the need for one. Recognising this need may result from:

- A Needs Assessment
- A Task Analysis
- Perception by people that a document is needed
- Result of a mishap that resulted in a loss of productivity or unsafe condition
- Addition of new equipment
- Implementation of new regulations; and
- Change in operating performance requirements.

3.3 Goal Definition

The goal of the document should be to address the user's information needs. This should be clearly defined, documented, understood, and agreed upon by any people involved in the generation or revision of the document.

3.4 Management Approval to Proceed

Creation/revision of instructional materials may require approval. Management responsibilities in document development may include:

- Approving the creation of a new document
- Requiring that instructional documents complement each other across the organisation by maintaining a consistent approach to document development

- Reviewing and approving proposed new instructional documents to verify the scope and intended content agree with existing documentation and management philosophy; and
- Requiring the initiator of the instructional material to justify its need.

For additional management-related information, refer to Chap. 6, "Managing Instructional Documents."

3.5 Information Gathering

Information needs to be collected and processed to produce a final document that contains the technically relevant and necessary information to perform the task(s). This information is usually collected by a group of people, including individuals considered to be subject matter experts (SMEs) and also those who are users of the document brought together for the purpose of developing the instructional materials. The information gathered is considered source material or supporting documentation.

3.5.1 Subject Matter Experts (SME)

A subject matter expert (SME) is an individual with a high level of knowledge in a particular job discipline or operation. They may come from inside or outside of the organisation. They can assist the document writer in defining complex processes and verifying and validating instructions during development.

3.5.2 Source/Supporting Documentation

Resource materials and other supporting documents are required to develop instructional materials. Often by the time document development begins, drawings and other technical documentation have already been generated and should be accumulated. Some of these resource materials are listed below:

- Related Vendor Documentation
- Related Drawings
- Relevant Regulatory Documents; and
- Company Policy Documents.

3.5.3 Writing to a Specific Audience

An important consideration in planning, writing, reviewing, and revising instructional materials is to know who the 'end user' will be, who will be reading and performing the document's steps. Some broad categories of audiences include:

- *Non-specialists*: Those with the least technical knowledge
- *Technicians/Operators*: Those who operate, maintain, and repair equipment
- *Experts*: Those who have in-depth knowledge of the theory and operation of the equipment; and
- *Executives*: Those who make business, economic, administrative, and other decisions.

For example, if technicians are the user, the document should be directed to technicians only. Generic information can be directed to multiple audiences, but technical information should be directed to a specific audience. Writers should identify the person expected to perform the individual steps, such as "Console Operator" and "Crane Operator." This is necessary so that people understand which steps they are responsible for from the start to finish of a process or task. Although English can be considered the universal language to be used at sea, for many people, English is a second language and may pose difficulties when instructional materials are developed only in English. It is appropriate for instructional materials to be translated into the native languages of the users provided the documents go through a verification and validation that the intent of the instruction is complete, correct, and appropriate. Refer to Appendix D, "Examples of Instructional Materials", for sample procedural steps that have been translated from one language to another.

3.6 Effective Presentation of Material

Documents that are too long, too detailed, irrelevant, disorganised, poorly explained, or difficult to access can lead to inefficiency and higher error rates, resulting in damage to equipment and/or injury to personnel. To help avoid this, a systematic approach can or should be used to present the information. This is outlined in the text below. A systematic approach to developing instructional material is recommended to:

- Promote a thorough-document development process
- Meet the user's needs
- Meet the needs of the Company; and
- Effectively present the contents of the document.

A systematic approach used by writers should include the following elements:

- *Analysis*—Assessing compliance with regulatory and company requirements
- *Design*—Includes designing implementation plans, and identifying both required and needed materials
- *Development*—Includes the actual development of instructions and related training information
- *Consultation*—Includes gathering feedback regarding the effectiveness of the document with the end user
- *Evaluation*—Evaluating the effectiveness of the document and the document development approach; and
- *Implementation*—Includes implementing instructional materials in the field including training personnel on the process.

3.7 Enhancing Usability

The usability of instructional materials is a product of the information they present, the ease in which the information can be located and understood, word clarity, step clarity, aids to decision-making, etc. Usability is increased through user participation in the document development process, good design, and a document control system.

3.7.1 User Participation

It is important that users of the materials actively participate throughout the development process. The users who are most likely to benefit from the document development process are the experienced persons (e.g., subject matter experts). Their involvement should include developing a task list, determining document needs, selecting an appropriate format including input into the writing, reviewing, and maintenance of processes.

3.7.2 Instructional Material Effectiveness

Document effectiveness is the result of a good design and development process, with this practice adopted by the owner, operator, or manufacturer and repeated for each task requiring instructional material development.

3.7.3 Instructional Material Maintenance

Document development is a continuous process because instructional materials can become outdated and should be maintained regularly and consistently to reflect current practices or reflect current system design. For this reason, a control system should be implemented. All instructional materials in a system should be tracked, reviewed periodically, and disposed of when no longer relevant. (Refer to Chap. 6, "Managing Instructional Documents".)

3.8 Instructional Material Writing Elements

It is important that the lists of instructions in documents describing the process by which a task is completed are user intuitive. Technical manuals, for example are normally developed by a vendor or manufacturer and not developed "in-house." These instructions are highly analogous to procedures and subsequently should follow the guidance in the following paragraphs. The writing elements listed in the Paragraphs below will help writers develop effective instructional materials.

3.8.1 Instruction Number

The instruction number should be clearly identified, taking care to number instructions in the order in which they are intended to be performed. For instructions which are lengthy or complicated, they may need to be broken down into smaller steps, making sure these are also numbered appropriately. Consider the use of numbers, roman numerals, or letters for the different levels of instruction.

3.8.2 Clarity

Clarity is a prime consideration. To be effective, instructional materials should be perceived as clear and unambiguous. This requires the elimination of imprecise and confusing wording. It also requires that information be conveyed as explicitly as possible to remove any doubt as to what is required.

Note: *It is important that documents remain comprehendible if being translated from one language to another language.*

Literal translations often can lead to instructions that do not make sense. Refer to Appendix D, "Examples of Instructional Materials", for examples of documents that were translated from one language to another.

3.8.3 Brevity

Brevity involves using a limited amount of words to express an idea by being short and specific. Instructional documents should be constructed using brief sentences within individual steps, notes, cautions, and warnings. This involves reducing the number of words needed to express what is intended while maintaining clarity.

3.8.4 Use of Imperative (Command) Sentence Structure

The active voice and present verb tense should be used in conjunction with the imperative tone, which is a grammatical tone that expresses the will to influence the action of another person (i.e., a command). The instructional materials should be developed in a manner that directs the person to perform a specific action and not be suggestive. Sentences are considered in the active voice when the subject of the sentence performs the action. The present tense of a verb (action word) shows that an action is taking place in the present.

Example:
Set up the test box on the power panel board.

3.8.5 User Friendliness/Usability

User friendliness is a user-focused information presentation style that makes complex information easy to access, read, use, and understand. This includes:

- Writing in a clear, concise, organised manner to reduce complexity and eliminate ambiguity
- Using an information presentation style that makes it easy for a user to find the necessary information (includes navigational aids for online documentation)
- Presenting the information in a style that is pleasing to the eye
- Sequencing the information logically
- Consistent ordering and placement of same-type information
- Including necessary information
- Delivering information in a format that meets the user's need; and
- Making the information available to the user in a timely manner.

3.8.6 Technical Accuracy

Technical accuracy is near perfect conformity to fact or truth. If the instructional material is not technically accurate, the consequences can be severe. User expectation can be compromised, and information may be lacking to perform the task(s) associated with the document in a safe and thorough manner.

3.9 Style Format Guidelines

Format refers to the shape, size, and general layout of a document. Layout refers to the arrangement of text and graphics on the various pages. The term "format" is often used to describe both style and format issues. The guidance in the following paragraphs is effective for good instructional material development.

3.9.1 Font

Instructional documents should be developed using Sans Serif font to aid in visual clarity. Standard font size for instructional materials should be at least 12 pt.

3.9.2 Sentence Structure

3.9.2.1 Simple Sentences
Use simple sentence structures and exclude unnecessary descriptive words. Including too many nouns and adjectives detracts from the point being made and can make a simple piece of information unclear. Write instructional materials using simple command statements. This can be done by omitting unnecessary articles (the, an, a).

Example:

Prefer:
   Close and secure manway hatch at second level. (12 pt. font)

Not this:
   Close and secure the manway hatch at the second level (10.5 font)

3.9.2.2 Imperative Tone
Use imperative (command) sentences. Write instructional steps as action-oriented sentences, avoiding passive statements.

Example:

Prefer:
   Open valve V-263.

Not this:
  Valve V-263 is to be opened.

3.9.2.3 Avoiding Negative Sentences
Avoid negative sentences. Write instructional steps using positive statements.

Example:

Prefer:
  Close enclosure door.

Not this:
  Do not leave enclosure door open.

3.9.2.4 Keeping Thoughts Related
Avoid unrelated thoughts in one sentence. Use only one main thought or action per sentence.

Example:

Prefer:
  Step 1. Verify tools have been removed
  Step 2. Clean any spills
  Step 3. Hold coordination meeting with involved personnel to discuss startup of compressor

Not this:
  Verify tools have been removed and any spills have been cleaned. Hold coordination meeting with involved personnel to discuss startup of compressor. The third sentence in the above example should be in a separate step.

3.9.2.5 Length of Sentences
Keep sentences short. Where possible, to assist with instructions being concise and readable, assume that more than twenty (20) words is too long. Ten (10) words or fewer are preferred.

3.9.2.6 Listing Multiple Objects
List multiple objects separately. Use a numbered or bulleted list, rather than paragraph format. A numbered list indicates that the actions must be performed in the sequence shown. A bulleted list indicates that the actions can be performed in any sequence.

Example:

Prefer:
  Open the valves in the following order:

3.9 Style Format Guidelines

1. V-103
2. V-104
3. V-107
4. V-213
5. V-222

Not this:

Open the following valves: V-103, V-104, V-107, V-213, and V-222.

3.9.2.7 Multiple Alternatives

List multiple alternatives. Do not write long, unbroken sentences that contain too many ANDs and ORs. Place the alternatives in a bulleted or numbered list.

Example:

Prefer:

Step 1. Switch evaporator mist eliminator wash to AUTO.

Step 2. If distillate conductivity rises without explanation (such as decrease in sump pH), increase duration or frequency of mist eliminator wash cycles until distillate conductivity returns to normal.

Step 3. As required, use globe valve V-107 to adjust mist eliminator wash supply so that pressure stays at 2 psi during wash cycle.

Not this:

Switch evaporator mist eliminator wash to AUTO and if distillate conductivity rises without explanation (such as decrease in sump pH), increase duration or frequency of mist eliminator wash cycles until distillate conductivity returns to normal, then as required, use globe valve to adjust mist eliminator wash supply so that pressure stays at 2 psi during wash cycle.

3.9.2.8 Conditional Statements

State conditions before actions. In writing conditional statements, write the condition (the "if," "after," or "when" statements) as the first clause and the contingency (the "then" statement) as the second clause.

Example:

Prefer:

After closing drain plug, fill tank with approximately 38 °C water, add detergent, and then press AGITATE button on control panel.

Not this:

Fill tank with approximately 38 °C water, add detergent, and press AGITATE button on control panel after closing drain plug.

If/Then tables can be used to express multiple conditional statements and actions for the user to take in each condition.

Example:

If...	Then...
Battery voltage is less than 125 VDC	Record voltage value in the provided table
Battery voltage is greater than 125 VDC	Do not record voltage value in the provided table

3.9.2.9 Abbreviations and Acronyms

Spell out abbreviations and acronyms the first time they are used to clarify the meaning of an abbreviation or acronym. Use of an abbreviation or acronym is then enclosed in parenthesis. After the first appearance in a document, it is acceptable to use only the abbreviation or acronym.

Example:

Prefer:
   International Maritime Organisation (IMO)

Not this:
   IMO. (As a first-time appearance)

3.9.3 Vocabulary

Document writers should use vocabulary words and terms that are easily read and understood by the end users. The following guidelines should be observed:

3.9.3.1 Consistent Terminology

Be consistent with the use of terminology. Equipment, for example, should be referred to using the same word or term throughout the document so that the user has no doubt about which piece of equipment is being referred to. Where alpha-numeric are used as equipment designators, place the numeric designation first, then the alpha name (or vice versa, according to the company or manufacturer's adopted convention).

Example:
   C-292 Booster Compressor

Example:
   Booster Compressor C-292

3.9.3.2 Specificity

Use specific words that precisely describe the task or action. Avoid ambiguous and vague instructions such as "check frequently" or "adjust slowly." Where possible, use specific intervals or guidelines, for example: "check every four (4) hours", or "adjust by performing one (1) rotation per second". To assist with specificity, a Constrained Word List is provided in Appendix B, "Example Constrained Word List and Constrained Language List". A Constrained Word List is a list of preferred words and terms shown adjacent to corresponding non-preferred words and terms. This list should also include company-specific terminology. Document writers are encouraged to use the Constrained Word List for preciseness.

3.9.3.3 Use of "Check," "Evaluate" and "Verify"

When using the terms "check," "evaluate," and "verify," abide by the following definitions to choose the appropriate term:

Check

To observe an expected condition or characteristic; to determine; to ascertain.
Example:
"Check pressure gauge for pressure greater than 45 psi

Evaluate

To assess; to determine the importance of; to appraise a situation (implies technical knowledge).
Example:
"Evaluate current pump pressure."

Verify

(1) To make sure by taking necessary or appropriate actions.
Example:
"Verify discharge pressure does not exceed maximum levels."
(2) To establish the truth or accuracy of.
Example:
"Verify readings before recording them."

3.9.3.4 Use of "May," "Should," "Will," "Shall," and "Must"

Commonly accepted definitions for the following words, "may," "shall," "will," "should," and "must" are provided below:

- *May*: Indicates acceptable or suggested methods. Denotes permission (neither a requirement nor a recommendation)
- *Should*: Indicates recommended or preferred method

- *Will*: Indicates a requirement
- *Shall*: Indicates a requirement; and
- *Must:* Indicates a requirement, used where a provision is a regulatory requirement.

3.9.4 Level of Detail

Document writers and SMEs sometimes disagree regarding the level of detail that should be included in a document. As a result, there can be a trade-off between too much or too little detail. Too much detail will waste resources in writing and may increase the time required for task performance. Too little detail may result in the task being performed in an unsatisfactory or unsafe manner.

3.9.4.1 Considering the User and Environment
The level of detail should be consistent with the knowledge, skill level, and training of the persons who will perform the task. While developing instructional materials, consideration should be given to varying conditions (temperature, noise, lighting, etc.) where the task will be performed. At times, a person may perform a task in a stress-free environment, while at other times the same task may be performed under conditions of stress and extreme environmental conditions. Either condition should be accounted for in the document.

3.9.4.2 A Guideline for Level of Detail
In consideration of the above, the appropriate amount of detail to be included in a document can be viewed as the amount of detail sufficient for a person to be trained to the skill level necessary for the specific task. A person should be given the information to perform the task even if that person has never performed the task before.

3.10 Document Organisation

Instructional materials should be organised logically so that steps or instructions are carried out in a manner that is expeditious and safe. The logical relationships of information associated with a task should be defined clearly. Without a logical organisation, personnel cannot be expected to correctly perform the given task. Write a general outline, detailing only the main sections or parts of the document, maintaining the proper order. Consider using broad categories during this phase, as a reference for developing the general outline. As the outline starts taking shape, more specific categories can be developed. The remainder of the organised information should be broken down further into separate subsections and paragraphs to formulate general ideas within each main section. Create descriptive titles for these subsections in the outline to promote organisation and ease of reference.

Once this has been completed, the process of developing a document will become easier and efficient for the writer as there is now a working platform. Figure 3.1, "Sample Template", which is at the end of this chapter, outlines each of the elements of a document as described in the following paragraphs. The following paragraphs provide suggestions for the organisation of instructional materials.

3.10.1 Front Matter

The information in the front part of the document is called the Front Matter. This Front Matter information is not necessarily required and can differ according to the needs of the owner/manager, or manufacturer. Generic elements of the Front Matter are discussed below for guidance:

3.10.1.1 Basic Front Matter Elements

(1) *Title*: The title of the document should be clear and concise. It should include the name of the equipment involved and the objective of the document. To keep the title concise, the length of the title should be kept to 10 words or less. Key action words should be positioned at the beginning of the title.
   Example:
   "Shut Down Air Compressor for Maintenance."
   Example:
   "Shutting Down Air Compressor for Maintenance"
(2) *Procedure Number*: The procedure number should be clearly identified. For ease of recognition, procedure numbers can include both an alpha and a numeric designator. The alpha portion can designate the type of procedure while the numeric portion can designate the sequence. This should be considered as applying only to procedures.
   Example:
   MP078 (for Maintenance Procedure 078)
   Example:
   EP016 (for Emergency Procedure 016)
(3) *Date*: The date should be prominent and displayed in a format that excludes confusion as to which day, month, or year is being referred. Example: 09 November 2009. The date when the document was developed may be different than the displayed revision date (refer to "Revision Date" below)
(4) *Revision Date*: The date of the last revision should be clearly displayed on the front page of the document. The revision date helps personnel to identify a current or out-of-date materials

Title: LOCKOUT/TAGOUT PROCEDURE			Date:	← Front Matter
Procedure #: **1234.567**			Revision #: **2**	
Facility: Main	Area: Engine Room	Description: Compressor	Equipment #	

Person Responsible: John Smith	← Person Responsible/Initiator Statement

Purpose:	← Purpose Statement

CAUTION
Hazardous chemicals present. Wear appropriate PPE and verify ventilation is adequate prior to servicing.   ← Non-Step Info/Special Equipment

SHUTDOWN, LOCKOUT, TAGOUT, and TEST SEQUENCE

#	Step	Description
1	Notify	Notify all affected employees that servicing or maintenance is required on a machine or equipment and that the machine or equipment must be shut down and locked out to perform the servicing or maintenance
2	Review Lockout Procedure	The authorized employee shall refer to the company procedure to identify the type and magnitude of the energy that the machine or equipment uses, the hazards of the energy, and shall know the methods to control ← Procedure Steps
3	Perform Machine Stop	If the machine or equipment is operating, shut it down by the normal stopping procedure (depress stop button, open switch, close valve, etc.) A reference for the operating procedure for shutdown should be placed here.
4	Isolate Energy	Follow typical lockout-tagout procedure from top to bottom to deactivate the energy to isolating device(s) so that the machine or equipment is isolated from the energy source(s). Note: It may be necessary to dissipate the non-lockable energy sources before isolating the lockable energy sources. (i.e., lower machine to lowest position before locking out)

RESTORE TO SERVICE SEQUENCE   ← Summary Statement

#	Step	Description
1	Check Machine	Check the machine or equipment and the immediate area around the machine to verify that nonessential items have been removed and that the machine or equipment components are operationally intact.
2	Check Area	Check the work area to verify that all employees have been safely positioned or removed from the area
3	Verify Machine	Verify that the controls are in neutral
4	Remove Lockout	Remove the locks, tags, and lockout devices and re-energize the machine or equipment. Reverse the order of all lockout-tagout procedure steps from bottom to top starting from the last page. Note: The removal of some forms of blocking may require re-energizing of the machine before safe removal.

END OF PROCEDURE   ← End of Procedure

Procedures Prepared By: Procedures Authorized by:

Date: Date:

References: ← References

Fig. 3.1 Sample template

(5) *Revision Number*: The revision number can be used to determine if a document is current or out-of-date when a searchable database is available to determine its currency; and

(6) *Persons Responsible*: Some instructional materials include the name of the person who was responsible for creating the document and/or who has continuing responsibility for maintaining and updating when necessary.

3.10.1.2 Table of Contents

For documents that are long and complex (i.e., technical manuals), it is suggested to include a table of contents. The writer of a technical manual should include a table of contents right after the front matter (refer to this chapter, Sect. 3.10.1, "Front Matter"). The table of contents should accurately list the sections, subsections, appendices, and any relevant figures and tables within the document for easy use.

3.10.1.3 Initiator Statement(s)

An "initiator" statement describes the personnel (i.e., position, skill level, certifications obtained) who are authorised to perform the procedure or instruction. The advantage to specifying the initiators is that there should be no confusion regarding who is authorised to perform the task. Inclusion of such statements is optional and specific to organisations.

3.10.1.4 Purpose Statement (Scope)

The purpose statement should clearly identify the main objective(s) of the document. If the word "Purpose" or "Scope" is indented to the left, for example, in a separate table cell, the statement may begin with the word "To," omitting "The purpose of this document is_ _ _."

Examples:

Purpose: To establish a sequence of safe operations and actions to be taken if an emergency alarm signal is activated in the engine room.

Purpose: To supply steam and condensate to the equipment in Unit 200.

3.10.1.5 Prerequisites

Prerequisites include the required conditions that must exist before beginning the task or instructions and may include the required safety equipment the person must wear during performance of the task, as well as any tools that they will need, if this information is not put in a separate location such as "Safety" or "Required Personal Protective Equipment (PPE)."

3.10.1.6 References

References are documents that:

(1) The person will need during performance of the task; and
(2) Define the work that must be performed in accordance with the referenced document(s)

These can be company policy documents, regulatory documents, and other documents when the situation calls for performing another task.

Example:
SG (Safety Guidance) #54, Safe Welding Practices

Example:
SD-1224, Shut Down Fire Water Pump

When permits must be obtained prior to performance of a task, they should be listed on the first page of the document, showing the appropriate title and alpha-numeric designation, if used.

Example:
P-108, General Work Permit

Example:
P-115, Confined Space Entry Permit

3.10.1.7 Safety/Environmental Concerns

Any safety consideration that may be encountered during performance of the task should be identified on the front page of the document. Environmental concerns should also be identified, such as the possible accidental discharge of a chemical to the atmosphere.

(1) *PPE*. The required personal protective equipment that must be worn while performing the task can be identified here unless put in a separate section, such as "Required Personal Protective Equipment;" and
(2) *Possible Chemical Exposure*. Where the likelihood of exposure to chemicals in the work area exists, it is recommended that each chemical be listed along with the Material Safety Data Sheet (MSDS). Persons should have an awareness and understanding of material safety data for raw materials, intermediates, products, and effluent/waste.

3.10.1.8 Special Equipment Needed

If tools are needed to perform the task, they can or should be listed in this section. This is worthwhile if the task is to be performed in a remote location because failure to have the tools available can result in lost time and added costs.

3.10.2 Document Sections

The sections of instructional materials can be divided into Summary Statements (headings), Main Steps, Bulleted or Numbered Lists, and Notes, Dangers, Cautions, and Warnings.

3.10.2.1 Summary Statements (Headings)
For purposes of dividing the task into logical groups of steps, summary statements should precede their related steps as a heading. By doing so, lengthy tasks are divided into groups of steps, reducing confusion and helping the personnel to easily understand the various objectives of the task.

Example:
   Preparations for Centrifugal Pump Disassembly

Example:
   Disassemble Centrifugal Pump

Example:
   Reassemble Centrifugal Pump

3.10.2.2 Non-step Information
Non-step related information includes explanatory and cautionary information placed in Dangers, Cautions, Warnings, and Notes. Table 3.1, "Usage of Signal Words" below outlines the situations in which the signal words should be used, as well as where the information should be placed within an instructional document.

3.10.2.3 Task Steps
Task steps should be numbered sequentially for appropriate performance of the process. Only one action is performed per step. However, as many as three actions can be performed per step if they are closely related and performed simultaneously. Example: Open the enclosure, release the filter locking mechanism, and pull out the old filter. As noted below, the hierarchy of steps includes three or four levels, along with numbered and bulleted lists. Steps should start with an action word, except for conditional steps [refer to this chapter, Sect. 3.10.2.3(2)) below], using the active voice as much as possible.

Table 3.1 Usage of signal words

Signal word	When to use…	Placement in procedures…
Danger	In imminently hazardous situations that, if not avoided, will result in death or serious injury; to be limited to the most extreme situations	BEFORE the relevant step or procedure
Warning	In potentially hazardous situations, which, if not avoided, could result in death or serious injury; used to signify situations that could result in serious damage to vital equipment, or a major pollution problem	BEFORE the relevant step or procedure
Caution	In potentially hazardous situations which, if not avoided, may result in minor or major injury; used to signify situations where property damage or a minor pollution problem is possible	BEFORE the relevant step or procedure
Note	When extra information that helps explain a process as necessary, but may or may not be needed for completion of the task. Can include background information that may have a minor role in supporting decision making by the person but is not considered absolutely necessary to know for task performance	Before or after the relevant step

Example:
Step 3 Perform the following to test well at platform:

(1) Branching Steps. Branching steps refer to another step at a different location in the document. They direct the person to a previous step in the task but may branch to a later step, for example, when one or more steps can be skipped due to a specific condition. When completing the development of a document using branching steps, branches should be verified as correct prior to distributing the document for implementation.

Example:
Step 2.5 After cleaning tank T6, go back to Step 1.3 to begin cleaning tank T7.

(2) Conditional Steps. Conditional statements placed within a procedure step describe a condition or decision that must occur before an action is performed. If the action is performed in accordance with the condition, the action can correct or avoid deviation. When writing conditional steps:
- Clearly identify the condition or the basis for decision-making
- Put the WHEN or IF before the THEN
- Adequately address decision paths; and
- If there are three or more conditions, they should be listed vertically beneath the associated step.

Example:

When lab report shows no harmful bacteria found, place potable water tank in service.

3.10.3 Ending the Procedure

The end of a procedure or series of steps should be clearly identified. The simple statement, "End of Procedure" is all that is needed and should be placed after the numbered steps and before any appended information.

Example:

End of Procedure

3.10.4 Appended Information

Information appended to a document includes additional information such as drawings, schematic diagrams, valve lists, parts lists, and equipment lists. This information is needed for performance of the task. Information related to the actions but not directly used in the actual performance of the individual steps is listed in the Reference section at the end of the document.

Graphics in Instructional Materials 4

4.1 General

Some instructional materials can be detailed and describe more complicated or involved tasks than one single set of instructions. Graphics and pictorials are non-text materials that support a document, such as figures, checklists, tables, and data sheets. The use of graphics or any other visual aid can help clarify a specific task for the user and should complement the text, providing an alternative and often more intuitive way of presenting the information. Illustrations and diagrams of equipment and machinery are of particular importance. When describing a process for repairing or maintaining an overly complex piece of machinery or equipment, detailed drawings are helpful for the user. Additionally, an instructional document may be used and interpreted by many nationalities throughout the life of the equipment, task, or document. Therefore, graphics and visual images should benefit users from different cultural backgrounds to help identify or understand the presented material. It is important to remember the use of graphics, pictorials, or other media to represent an idea. It should help explain the ideas that are contained in the document and presented simultaneously with the associated text. Graphics are a good supplement to text and may also help keep users interested. Always use high-quality, relevant graphics to communicate to the audience effectively.

4.1.1 Definitions

Diagram: An illustration that shows the parts of something or how it works. A diagram has labels that identify the names of various parts with the use of captions.

Chart: Contains important facts about a subject that are arranged in a way that makes them easy to see and understand.

Concept: A fundamental category of existence derived or inferred from specific instances or occurrences.

Flowchart: A graphical way to present the logical steps in the decision-making process. This provides the user with an easy-to-follow mechanism though a series of logical decisions and the steps that should be taken as a result.

Graphic: Communication of information through pictures, figures, data, and sometimes words.

Illustrations: Used to help the user understand the information being presented and can include photographs, drawings, diagrams, maps, flowcharts, and many more.

Pictorial: An illustration or picture within a piece of writing. Pictorials should be used within instructional materials to promote full understanding of instructions.

Tables: An efficient way to present large amounts of quantitative information in a compact space in rows and columns.

4.1.2 Graphics

Graphics can be used to represent objects, words, concepts, or numbers. Graphics are also used to show the key components of a task or document users are working with. This could include before and after views, close-up views, or detailed illustrations of key actions that users must perform. Some general criteria guidelines are as follows:

(1) Graphics should be clearly labelled and easily identifiable
(2) Graphics should be appropriate for the intended use
(3) Graphics should be placed next to corresponding text and clearly labelled
(4) Separate graphics should be used for each distinct idea
(5) Graphics should be large enough to see the focal point and key words clearly; and
(6) Graphics should be referenced properly and in order within the document.

4.1.3 Sources of Graphics

Once the need for a graphic has been identified, it is important to locate or create the graphic that will act as a supplement to the user to better understand the information in the document. A graphic can be developed from scratch or borrowed from another source, depending on the process to be explained. The amount of details included in a graphic can range from minimal (i.e., simple line drawing of a single object) to maximal details (i.e., 3-dimensional diagram depicting a complex process) depending on the ultimate goal of the graphic. Graphics may already be available in other documents or on the internet. It is important to search for already existing documents before creating a new graphic and reference the source of the graphic if it is borrowed and obtain authorisation in accordance with local copyright laws.

4.1.4 Types of Graphics

The term "graphics" includes many types of visual aid materials that support instructional documents or text. Using graphics breaks long processes into shorter sub-processes that consist of only a few steps. The following are several types of graphics that may be used to create instructional documents:

- *Figures* include all types of graphics other than tables, including charts, maps, and photos
- *Graphs* represent pictorially the relationship between variables, as the variables increase or decrease. Graphs are useful for showing trends and comparisons with the standard or ideal curve
- *Illustrations* can be especially useful to indicate an item's location, for example, a pushbutton on a panel or the packaging around a valve. Instructional documents that deal with equipment maintenance usually require illustrations to show, for example, how parts fit together in a larger assembly or the internal parts of a piece of equipment. Some common types of illustrations include sketches, exploded view, cutaway, schematic, and sample forms
- *Tables* only contain words and numbers. Tables are valuable to instructional documents because they enable the writer to present a large amount of data in a small amount of space
- A *technical drawing* is a common type of figure that is used in instructional documents as it represents a detailed or complicated process. It is important to clearly label and identify the different parts of a technical drawing to provide clarity to the user; and
- *Charts* are figures that display data in an organised visual format that is easy for the user to understand.

When the writer of an instructional document has supporting material that is pictorial rather than numerical, writers should choose photographs, drawings, and diagrams as their graphic aids. Pictorials should be a clear representation of a detailed principle that supports the text. The main point of the pictorial should be obvious to the user. This may be achieved through good lighting, good angles, and an absence of irrelevant background detail in a photograph. The following are examples of different types of graphics used in instructional materials:

Figure 4.1 is an example of a diagram that shows the human body in three (3) translational axes. Although this diagram is 3-dimensional, it is simplified and clearly labelled so the user is easily able to understand the important aspects of the diagram.

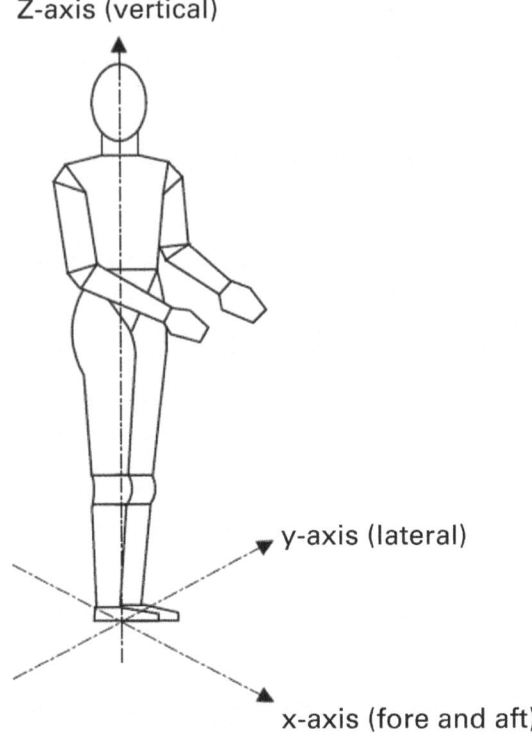

Fig. 4.1 Three (3) dimensional diagram

Figure 4.2 is an example of a flowchart. This flowchart clearly shows the whole process for collecting and analysing vibration data. The flowchart outlines the critical steps from beginning to end and shows how to proceed when each step has been completed.

4.1.5 Guidelines for Using Graphics in Instructional Materials

It is possible to misuse graphics. The following are some tips to verify that graphics are used correctly and are relevant to the information in the document.

(1) Avoid placing graphics in confusing places or sequences
(2) Graphs, charts, or any other visual representation of the data should be properly labelled, or colour coded. Use symbols or colours that are easily distinguishable. The user should be able to understand the graph or chart with a quick glance. For example, a pie chart must move in the clockwise direction, and each section of the pie must be properly labelled
(3) Use concise figure titles for graphics
(4) Discuss the graphic in the text somewhere close and relevant to the graphic

4.1 General

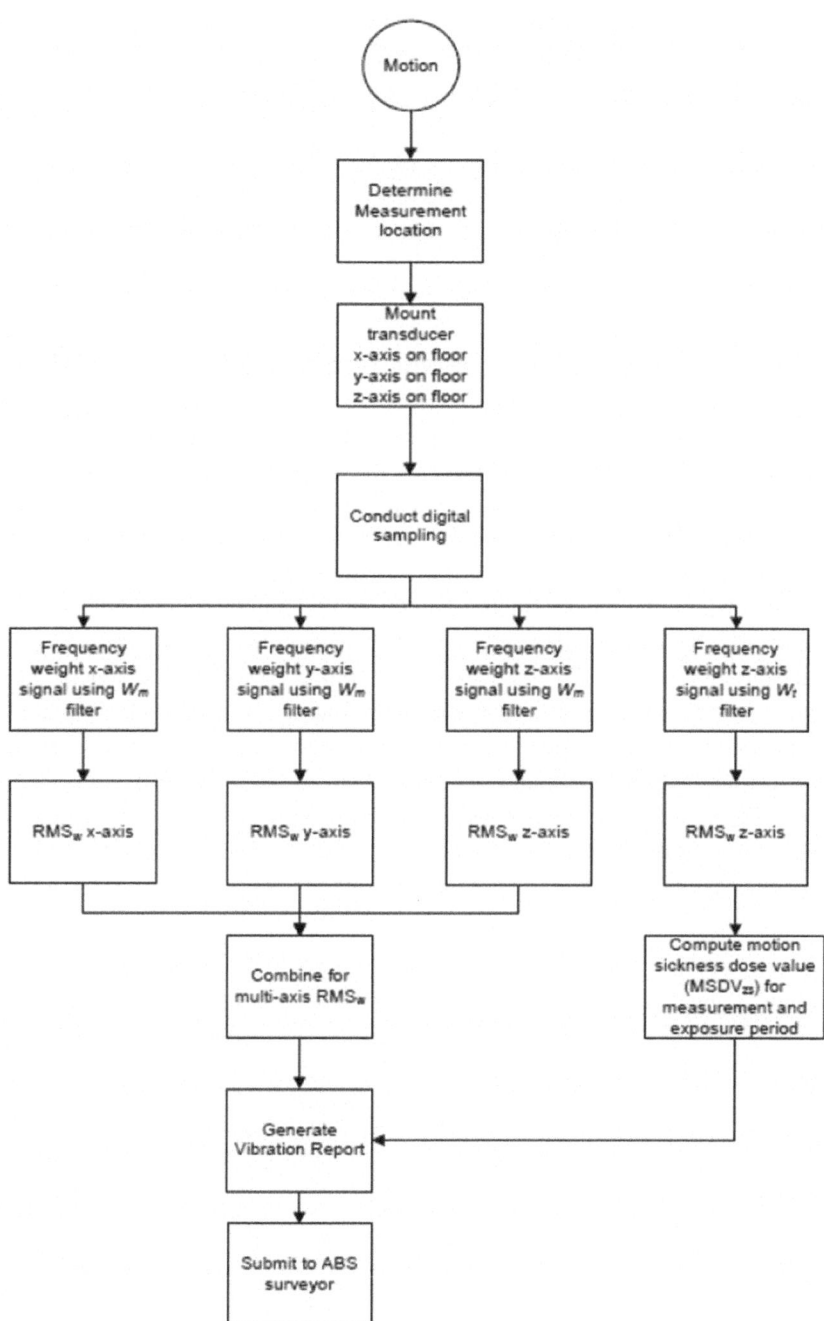

Fig. 4.2 Flowchart showing a process

(5) The graphic should be appropriate for the subject matter and audience. The end-user should be able to easily understand the point of the graphic
(6) Obtain necessary permissions and include the source of any graphic that is borrowed, including tables, graphs, illustrations, charts, etc.
(7) Include graphics and text on the same page. Additionally, graphics should not be placed on pages by themselves or as an attachment at the end of a document; and
(8) The graphics should fit within the normal margins of the document, if they do not, the graphic should be properly cropped or edited without distorting the image.

4.1.6 Incorporating Electronic Media into Instructional Materials

Instructional materials are not only developed and available in paper format anymore. Web-based media has become quite common for many types of instructional documents. Electronic filing systems and electronic media help companies organise and distribute important records and documents. It is time and cost efficient to switch to a digital filing system. It is easier to file older versions of instructional materials, current documents and to provide the user with any informational updates to documents. Another benefit of an electronic system is that this information can be accessed from anywhere with access to a computer through the use of a cloud storage or similar. Smartphones, tablets, and other technologies have made it quite easy to access the internet or a network from anywhere. This is beneficial for manufacturers, owners, and operators, as it is possible to digitise the necessary instructional materials and distribute them to the users no matter where they are located. There are also computer and smartphone apps that can be used to share and access documents between multiple users. Using digitised documents also reduces the chances of misplacing important notes or documents.

5. Verifying, Validating, Approving, Certifying, and Implementing Instructional Materials

5.1 General

This chapter assumes that a draft copy of a new/revised instructional document has been completed. The document should then go through the following processes before being implemented: verification and validation, approval, and certification. These processes may not be necessary for all instructional materials. A new/revised document should go through two (2) review processes; first for clarity, continuity, and technical content, and second for accuracy and feasibility.

5.1.1 Definitions

Certification: The process through which a certifying organisation grants recognition to an individual, organisation, process, service, or product that meets certain established criteria or special qualifications.

Implementation: The process of moving an instructional document from concept to reality; the execution of a plan to have a new document distributed so that specific future work performance is in accordance with the correct procedural document.

Validation: The process of checking an instructional document to determine if it satisfies technical and functional requirements, and that it is usable and suitable for those responsible for using the instructional document.

Verification: Evidence that establishes or confirms the accuracy and completeness of the instructional document.

5.1.2 Final Phase of Instructional Document Development Process

The final phase in the document development process involves some or all of the following activities:

(1) Verifying and validating the document content for technical correctness and ease of understanding
(2) Resolving and incorporating final SME review comments
(3) Proofreading and final review for format and style
(4) Approving the document after it has been determined ready for implementation; and
(5) Implementing the materials.

5.2 Verifying and Validating Instructional Materials

Prior to being distributed for implementation, an instructional document should be verified and then validated.

5.2.1 Verification

Verification is accomplished by a final review of the technical content and accuracy of the materials. A verification of the document should confirm that:

- Steps are complete and comprehensive
- Steps are correctly ordered
- Danger, cautions, warnings, and related guidance are present and based on SME knowledge, risk and hazard assessments, and/or task analysis data
- Components to be manipulated are appropriately identified by labels or other means, and where components are difficult to locate, the document provides location guidance
- Technical limits are accurately presented (e.g., "Tank pressure should not exceed 300 psi); and
- Operating environment is addressed (e.g., presence of spray, ice, oily surfaces, or excessive noise).

5.2.2 Validation

Validation is the process of checking a document to establish that it satisfies technical, usability (ease of use), and functional requirements. This is typically achieved through a comprehensive SME and user walkthrough of the document in the work area, or using relevant drawings when a walkthrough assessment is not possible. The persons performing the validation via a walkthrough should look for and assess:

- Accurate and appropriate labelling/identification of components within the document (e.g., "Isolate Sump Pump 3A using valves ISO 23 and ISO 24")
- Opportunities to simplify instructional steps (e.g., if the logical progression of the task allows it, direct the user to perform steps in one location before moving to the next location, rather than requiring the user to go back and forth)
- Appropriateness of Dangers, Warnings, and Cautions used in the documents
- Technical inconsistencies (e.g., where a procedure refers to a unit as "US Gallons" and associated displays are in Liters not Gallons)
- Completeness of instructional steps (e.g., rather than "Isolate Sump Pump 3A" provide "Isolate Sump Pump 3A using valves ISO 23 and ISO 24"); and
- Ease of access (e.g., for a difficult to reach valve in an overhead location, the document should advise that appropriate access aids should be brought to the area, and for access to a normally locked area, the instructional document should advise to bring keys).

Any issues or concerns noted during verification and validation should be reviewed and assessed, and where deemed necessary, result in modification of the document in question.

5.3   Resolving and Incorporating Final Comments

Comments should be incorporated into the final version of the document. This usually involves the following activities:

- *Correcting Errors*: This is the easiest activity to complete. It involves correcting simple errors such as equipment names, and tag numbers; and
- *Improving Word Choices and Sentence Structure*: The purpose of this activity is to align the selected words with industry terminology, making them clear and easy to understand (refer to Appendix B, "Example Constrained Word List and Constrained Language List").

5.4   Proofreading and Checking for Format and Style

After the material has been developed to the final draft stage, verified and validated, it should be proofread, checked for consistency in format and style, and technically reviewed for clarity and conformity to the original intent of the document.

5.4.1 Proofreading

At this stage, the document should be proofread for clarity, spelling, and grammatical and nomenclature correctness. It is generally advantageous to have someone other than the writer proofread the document.

5.4.2 Instructional Document Format and Style

All instructional materials should be reviewed to verify the format and style guidelines are followed as noted in Chap. 3, "Writing Instructional Documents", or as set out and approved by the owner, manufacturer, or operator.

5.5 Approving the New/Revised Instructional Materials

After all final comments have been resolved; the documents should be presented to management for approval. Management should review the materials for alignment with the purpose of the document and the goals and objectives of the organisation or manufacturer. The finalised document should be approved as described in the organisation's Quality Management Plan or other approval process. The immediate supervisor, group of participating SMEs, and the organisation's quality assurance officer review and approve each document.

5.6 Certifying the New Instructional Materials

Some instructional materials may be required to meet certification guidelines set forth by outside agencies or regulatory bodies to comply with various laws, regulations, and standards. These agencies may perform periodic audits of instructional documents for compliance. For this reason, new materials that fall under the umbrella of these regulators should be developed with the specific certification criteria in mind. This helps prevent non-compliances and the resulting required rework to bring the documents into compliance.

5.7 Implementing New/Revised Instructional Materials

Implementation of new instructional materials can be managed in stages, such as planning, notification, and distribution. By taking a staged approach to implementing the change, both personnel and management are given time to understand how the change may affect

5.7 Implementing New/Revised Instructional Materials 45

the organisation and adjust accordingly. It is important that personnel are clear as to when the old document (if applicable) is no longer valid, and when the new document should be used.

5.7.1 Planning for Implementation

Personnel should be exposed to new or revised materials prior to implementation. This can take the form of workshops, training classes or, simply, distribution of the documents to the affected personnel. This gives them many opportunities to familiarise themselves with the changes or new information and to ask questions or to identify potential implementation problems prior to implementation. Persons affected by the new or revised materials should be given advance notice of its implementation. The early notification serves to identify potential problems and provides an opportunity for the affected persons to ask questions and provide feedback.

5.7.2 Distribution

Copies of new and revised documents should be made available to persons affected by the changes, whether in hard copy or electronic form.

Managing Instructional Materials

6.1 General

Managing changes to instructional material helps to prevent the introduction of new hazards or increased risk of existing hazards. It also allows for documents to be kept up-to-date while making sure that affected personnel are notified of changes. Once the documents have been developed or revised and approved, they must be implemented. Furthermore, they must be maintained for accuracy and completeness. Key aspects to effectively implementing instructional materials include document control, access, and training. Maintenance of instructional documents includes management commitment, management of change, and periodic reviews. A management process for instructional material changes should be in place which includes personnel assignments, administrative systems, and resource allocations. Implementing these processes makes periodic document evaluations and revisions a recognised part of business operations. An onboard fleet management system may aid in the confirmation of document control, configuration management, and maintenance and delivery of instructional materials. It is recommended that Manufacturers, Shipyards, and Owners take into consideration when and where these materials will be used. If they are to be used in conditions that could result in damage to the document, then appropriate precautions should be taken (i.e., working copies, durable placards, or laminated cards/pages).

6.1.1 Document Control

While the header in a document typically contains document control information, companies should have a process in place to verify that only approved, controlled documents are used. Management control of instructional materials should include the following:

- Inclusion of a date and/or revision number on all documents
- Keeping a log of who holds the original copy/version of the document
- Removal of out-of-date documents from the workplace and archiving any earlier versions; and
- Reviewing documents periodically to determine whether they should be updated.

If the system for managing instructional materials is not working, be prepared to change it.

6.1.2 Access

For people to use and accept the new or revised instructional materials, they must be readily accessible in a location that does not make it intimidating for the user to obtain the document (i.e., the users should not have to go to a supervisor's office to ask for a local hard copy). If the material is available electronically, then it must be easily accessible to all users. The easier it is to access the materials; the more likely personnel are to use them. When possible, instructional material should be provided in the work areas, or the users should be asked the best way for them to gain access and verify that this is sufficient. To promote accessibility, the following should be considered:

- Instructional materials should not be stored in a locked office or on computers that are not available at all times
- Any documents referenced in instructional materials should be readily available; and
- For proprietary or secure documents, management should invoke responsibility to provide/allow access to only those people who use those documents.

6.1.3 Training

If applicable, company training programs should be updated as instructional materials change. Using the actual instructional materials as part of the training program will familiarise workers with the content and format of the material.

6.2 Maintenances of Instructional Materials

6.2.1 Updates

For instructional materials to remain up-to-date and accurate, they need to be maintained. Instructional documents should be updated when a change occurs that affects operating methods or other information contained in the document. For example:

- New equipment is introduced
- New regulations put into place
- Reoccurring Incidents/Near Misses/User Requests; and
- Change in operating environment.

6.2.2 Periodic Reviews

Instructional materials should be periodically reviewed to determine currency and accuracy. Typically, reviews occur in one of two ways, either a cycle for review of all instructional documents is set (e.g., every three years), or reviews are event-based (changing based on lessons learned from some event).

Appendix A: Checklist for the Preparation of Instructional Materials

The following checklist is a summary of the steps discussed in this guidance. This checklist can be used by developers as an aid in the document development process. Not all steps in this checklist are necessary for every document or every organisation. It is up to individual organisations to determine the level of detail required by the instructional materials.

Step	Title	✓ or N/A	Comment
1	Has there been a request for a new or a modification for an existing document?		
2	Has management approved proceeding with the creation or update of the document?		
3	Has/have the goal(s) of the document been defined?		
4	Has the appropriate information been gathered?		
4a	Has a Risk Assessment been done?		
4b	Has a HAZOP study been done?		
4c	Has a HAZID study been done?		
4d	Has a Task Analysis been done?		
4e	Has a Discovery/Gap Analysis been done?		
4f	Has a Needs Assessment been conducted?		
5	Has gathered information been analysed?		
6	Has the design and layout of the document been determined?		
7	Is the development of the material complete?		
8	Has the document been verified?		
9	Has the document been validated?		
10	Has the document been approved/certified?		

(continued)

© The Editor(s) (if applicable) and The Author(s), under exclusive license to Springer Nature Switzerland AG 2025
A. A. Olsen and F. Karkori, *Development of Procedures and Technical Manuals for the Marine and Offshore Industries*, Synthesis Lectures on Ocean Systems Engineering, https://doi.org/10.1007/978-3-031-74863-9

(continued)

Step	Title	✓ or N/A	Comment
11	Has the document been implemented?		
12	Is there a process in place for managing changes to the instructional materials?		

Appendix B: Example Constrained Word List and Constrained Language List

This appendix has two (2) parts. The first is a Constrained Word List and the second is a Constrained Language List. The Constrained Words List provides appropriate words to be used when describing actions in an instructional document. The Constrained Language List provides a specific definition for those words from the Constrained Word List to be used in instructional materials. The intent is to use both lists to verify proper wording is used to describe all tasks. The first step is to determine the action being described, and then choose the appropriate word to use based on the definitions in the Constrained Language List. Neither list is to be considered comprehensive and should be updated/modified to include standard company terms and definitions.

Constrained Word List

The words in the Constrained Language List (in this appendix, part 2) are indexed here by general classification as in a thesaurus. When writing an instructional document (i.e., procedure step or technical manual instruction) dealing with a particular action, look in the corresponding section in this list for acceptable words to describe the action. Some words appear in several categories.

Execution actions	
Pack	Perform
Improve	Open
Close	Spin
Puncture	Prepare
Maintain	Repeat

Disengage	Support
Retract	Prevent
Manually initiate	Shut down
Engage	Unwind
Rotate	Provide
Manually trip	Start
Enter	Wrap
Shift	Purge
Match mark	Stop
Exit	Recall
Slide	Suspend
Monitor	Wait
Go to	Withdraw
Slip	
Miscellaneous actions	
Assist	Purge
Avoid	Recall
Back off	Recommend
Improve	Reduce
Maintain	Reject
Manually initiate	Release
Manually trip	Replace
Match mark	Request
Monitor	Rope off
Perform	Rotate
Prepare	Store
Prevent	Trace
Provide	Use
General physical actions	
Adapt	Hang
Adjust	Heat
Align	Insert
Balance	Lift
Bend	Load
Channel	Lock out

(continued)

(continued)

Clear	Open
Code	Rack in
Coil	Rack out
Conduct	Replace
Control	Request
Cool	Restore
Cycle	Rope off
De-energize	Rotate
Dispose	Shift
Eliminate	Store
Energize	Trace
Exceed	Trip
Extend	Use
Extinguish	Vent
Follow	
Guide	
Handle	
Breaker actions	
Activate	Shake
Align	Spill
Close	Stir
Pump actions	
Actuate	Pump
Cycle	Shift
Isolate	Transfer
Lock out	Trip
Operate	
Fluid actions	
Bleed	Immerse
Depressurise	Lubricate
Dilute	Pour
Dissolve	Pressurise
Drain	Pump
Draw in	Rinse
Dry	Spray

(continued)

(continued)

Empty	Squirt
Fill	Stir
Flush	

Electrical actions

Charge	Energise
Coil	Ground
Connect	Isolate
Crimp	Neutralise
De-energise	Regulate
Disconnect	Wire

Positioning actions

Centre	Position
Lower	Put
Move	Raise
Place	Set

Transference actions

Alternate	Remove
Give	Return
Interchange	Take
Obtain	Transfer
Receive	

Switch actions

Adjust	Position
Advance	Press
Depress	Pull
Drive	Push
Lock out	Rotate
Place	

Valve actions

Align	Operate
Close	Shift
Control	Throttle
Cycle	Transfer
Gag	Isolate
Isolate	Review

(continued)

(continued)

Line up	Select
Modulate	Specify
Open	Verify
Impelling actions	
Compress	Pull
Force	Push
Press	Tap
Gas actions	
Bleed	Empty
Blow	Fill
Choke	Inflate
Deflate	Pressurise
Depressurise	Vent
Mental actions	
Allow	Develop
Analyse	Diagnose
Calculate	Establish
Check	Evaluate
Compare	Identify
Correct	Initiate
Determine	Weigh
Sealing actions	
Cap	Plug
Close	Seal
Cover	
Agitation actions	
Mix	Tighten
Screw	Torque
Secure	Unlock
Tie	Unscrew
Documentation actions	
Complete	Number
Copy	Record
Date	Refer
Fold	Report

(continued)

(continued)

Go to	Sign
Implement	Submit
Initial	Tag
Mark	

Construction actions

Add	Distribute
Arrange	Include
Collect	Separate

Miscellaneous words

And	Rise
Can	Shall
Drop	Should
If	Spare
Locally	Standby
May	Then
Or	When
Per	Will

Visual actions

Inspect	Read
Light	Recognise
Locate	Scan
Observe	Show

Collective actions

Add	Distribute
Arrange	Include
Collect	Separate

Maintenance actions

Clean	Rinse
Coat	Scrape
Cut	Scrub
File	Notify
Grind	Trim
Paint	Wash
Patch	Wipe

Communication actions

(continued)

(continued)

Alert	Notify
Confer	Refer
Listen	Signal
Junction actions	
Apply	Connect
Attach	Disconnect
Clamp	
Measurement actions	
Add	Calibrate
Calculate	Estimate

Constrained Language List

This list identifies words which have specific meanings when used in instructional documentation.

PREFERRED WORDS are given in ALL CAPITAL LETTERS (however, they are not to be capitalised in documents, except in the cases specified elsewhere).

Common synonyms are given in lowercase letters along with references to the corresponding preferred words. These synonyms may be used as needed for clarity in certain cases.

Accomplish: See PERFORM.

ACTIVATE: To set in motion or make active. "Activate the turbines."

ACTUATE: To put into operation; to move to action. "Actuate ECCS."

ADAPT: To make fit a new situation or use, often by modifying. "Use bushing to adapt cap to valve outlet."

ADD: To put more in. "Add water to the battery."

ADJUST: 1. To bring to a specified position or state. "Adjust micrometre to given measurements." 2. To bring to a more satisfactory state; to manipulate controls, levers, linkages, etc., to return equipment from an out-of-tolerance condition to an in-tolerance condition; to readjust; to reset. "Adjust cable tension using turnbuckles."

Advise: See NOTIFY.

ADVANCE: To move forward; to move ahead. "Advance turbine speed."

Agitate: See SHAKE.

Aid: See ASSIST.

ALERT: To warn; to call to a state of readiness or watchfulness; to notify (a person) of an impending action. "Alert personnel that area will be cleared."

ALIGN: To bring into line, to line up; to bring into precise adjustment, correct relative position, or coincidence. "Align slot in turnbuckle barrel with slot in cable terminal."

Allocate: See DISTRIBUTE.

ALLOW: 1. To let; to permit; to give opportunity to. "Allow sediment to settle out." 2. To leave; to allot or provide for. "Allow a 2-inch slack in the rope."

ALTERNATE: To perform or cause to occur by turns or in succession. "Alternate between test instrument channels."

ANALYSE: To examine and interpret test or inspection results to determine system or equipment condition or capabilities. "Analyse generator inspection findings to determine need for repairs."

AND: Establishes that each of a series of actions must be performed with no alternatives available. Connects items and actions.

APPLY: To put; to lay or spread on. "Apply sealant to gap between access cover and equipment structure."

Note: Use "lubricate" rather than "apply lubricant."

ARRANGE: To order; to group according to quality, value, or other characteristics; to put in proper order; to organise. "Arrange components by size from smallest to largest."

Ascertain: See VERIFY.

ASSEMBLE: To construct; to fit and secure together the several parts of; to make or form by combining parts; to reassemble. "Assemble valve components in accordance with specified procedures."

Assess: See EVALUATE.

ASSIST: To give support or help; to aid. "Assist man B to lift the load."

ATTACH: To join or fasten. "Attach side plate (2) of assembly using 1/2-inch screws." Note: Use "tag" in preference to "attach a tag." Use "connect" in preference to "attach electrical leads."

AVOID: To keep away from; to prevent the occurrence of. "Avoid the use of excessive force in seating valve."

BACK OFF: To cause to go in reverse or backward. "Back off nut to nearest castellation."

BALANCE: To equalise in weight, height, number, or proportion. "Balance electrical loads on buses."

Be sure: See VERIFY.

BEND: To turn by force from straight or even to curved or angular, or to force, back to an original straight or even position. "Bend wire until it lies flat against turnbuckle wall."

BLEED: To extract or let out some or all of a contained substance. "Bleed off tank air pressure."

BLOW: To send forth air. "Check for obstructions by disconnecting hose at air inlet and blowing filtered air through it."

Break: See DISCONNECT. See REMOVE.

CALCULATE: To figure; to compute; to determine by arithmetic processes. "Calculate voltage in a circuit with 10 amp of current and 5 ohms of resistance."

CALIBRATE: To determine accuracy, deviation, or variation by special measurement or by comparison with a standard. "Calibrate torque handles at least once each month so that the accuracy can be depended upon."

CAN: Refers to a response.

CAP: To install caps; to provide with a covering; to install or provide with a device for closing off the end of a tube which has a male fitting; to recap. "Cap all lines which have exposed male fittings."

Categorise: See IDENTIFY. See SEPARATE. Note: For determining the classification of a supply item, use "identify."

CENTER: 1. To adjust so that axes coincide. "Centre bushing in opening." 2. To place in the middle of. "Centre pointer on dial."

Change: See REPLACE.

CHANNEL: 1. To form, cut, or wear a groove in. "Channel rods so that they can be inserted easily." 2. To direct fluid through a passage. "Channel flow into container A."

CHARGE: To restore the active materials in a storage battery by the passage of a direct current in the opposite direction to that of the discharge; to cycle. "Charge the battery for a short time before making a specific battery check."

CHECK: To observe an expected condition or characteristic; to determine; to ascertain. "Check termination criteria are being met."

Checkout: See TEST.

CHOKE: To enrich the fuel mixture of a motor by partially shutting off the air intake of the carburettor. "Choke engine as required to start."

CLAMP: To fasten or press two or more parts together so as to hold them firmly. "Clamp the tension meter to the cable by releasing the handle slowly."

Classify: See IDENTIFY. See SEPARATE.

CLEAN: To wash, scrub, or apply solvents to; to remove dirt, corrosion, or grease. "Clean all parts using non-abrasive cleaner."

CLEAR: To move people and/or objects away. "Clear the area."

CLOSE: 1. To block against entry or passage; to turn, push, or pull in the direction in which flow, or access is impeded. "Close access panel." 2. To stop flow (in valves). "Close valve." 3. To make an electrical connection to supply power (for electrical devices). "Close circuit breaker."

COAT: To cover or spread with a finishing, protecting layer. "Coat battery cables with grease."

CODE: To put into the form or symbols of a system used to represent words; to mark with identifying symbols. "Colour code equipment parts."

COIL: To make into the form or shape of a loop; to roll or twist into the shape of a circle or spiral. "Coil wire."

COLLECT: To bring together into one body or place; to accumulate. "Collect required hand tools."

COMPARE: To examine the character or quality of two or more items to discover resemblances or differences. "Compare readings from both instruments."

COMPLETE: To end; to finish; to accomplish specified procedural requirements. "Complete all requirements on data sheet before continuing."

Comply: See FOLLOW.

COMPRESS: To press or squeeze together. "Compress spring-loaded assembly until latch engages."

Compute: See CALCULATE.

CONDUCT: To lead, manage, or direct. "Conduct pre-work meeting." Note: Use "perform" in preference to "conduct" a test.

CONFER: To consult; to exchange views. "Confer with maintenance supervisor if necessary."

CONNECT: 1. To bring or fit together so as to form a unit; to couple keyed or matched equipment items; to attach, mate, or join; to reconnect. "Connect antenna cable to radio transmitter." 2. To attach or mate an electrical wiring connection; to plug in. "Connect VOM leads to test jacks."

Construct: See ASSEMBLE.

Contact: See NOTIFY. See SIGNAL.

CONTROL: To fix or adjust the time, amount, or rate of; to regulate or restrict. "Control electrical current generation and distribution."

COOL: To make or become lower in temperature. "Allow motor to cool before disassembling."

COPY: To make an imitation, transcript, or reproduction of. "Copy the procedure for filing."

CORRECT: To make or set right, to alter or adjust so as to bring to some standard or required condition. "Correct any error before proceeding with activity."

COVER: To protect or shelter by placing something over or around. "Cover valve internals when operator is removed."

CRIMP: To compress or deform a connection barrel around a cable to make an electrical connection. "Crimp a connector on the yellow wire."

CUT: To divide into parts using a sharp instrument such as a scissors or knife. "If prongs of cotter pin are too long, they should be cut to proper length."

CYCLE: To perform a process where the beginning and ending events or actions are the same. "Cycle pump."

DATE: To affix a date to. "Sign and date enclosure."

Decrease: See REDUCE, LOWER, or DROP. Do not use "decrease" because of oral communications problems.

DE-ENERGISE: To deprive of energy. "Confirm that the battery cables are de-energized."

DEFLATE: To release air or gas from. "Deflate shock strut to check fluid level."

Deliver: See TAKE. See SUBMIT. See DISTRIBUTE.

DEPRESS: To press or push down. "Depress pushbutton switch and then release."

DEPRESSURISE: To release gas or fluid pressure from. "Depressurise hydraulic system."

Destroy: See DISPOSE.

DETERMINE: To find; to investigate and decide; to discover by test, study, or experiment (implies technical knowledge). "Determine the amount of tension on cable by following specified procedures."

DEVELOP: To set forth or make clear by degrees or in detail. "Develop procedures fully."

DIAGNOSE: To recognise and identify the cause or nature of a condition, situation, or problem by examination or analysis. "Diagnose malfunction."

DILUTE: To make thinner or diminish the strength of by adding water or other fluids. "Dilute solution by using 1 part solution to 2 parts water."

DISASSEMBLE: To dismantle; to take to pieces; to take apart to the level of the smallest unit or down to all removable parts. "Disassemble valve bonnet."

DISCONNECT: 1. To sever the connection between; to separate keyed or matched equipment parts; to break. "Disconnect antenna cable from transmitter." 2. To detach or separate an electrical connection; to unplug. "Disconnect leads to power circuit."

DISENGAGE: To release or detach interlocking parts; to unfasten. "Disengage turning gear." Note: For circuit breakers, use "open."

Dismantle: See DISASSEMBLE.

DISPOSE: To get rid of; to destroy. "Dispose of unused hydraulic fluid left in the can."

DISSOLVE: To cause to pass into solution. "Dissolve mixture in 2 gallons of water."

DISTRIBUTE: To divide among several or many; to divide or separate, especially into kinds; to deliver. "Distribute copies of the procedure for review."

DRAIN: To draw off (liquid) gradually or completely. "Drain servicing hose after removing it from filter valve."

DRAW IN: To pull (liquid) up into a container through suction. "Fill hydrometer by drawing in electrolyte."

DRIVE: To move reactor control rods, either in or out. "Drive control rods into position 02."

DROP: Describes a decrease in a parameter as the result of an operator or equipment action. "Verify drop in pressuriser pressure."

DRY: To cause to be free from water or liquid. "Dry bearings with low-pressure air."

Effect: See PERFORM.

ELIMINATE: To expel; to ignore or set aside as unimportant. "Eliminate all unnecessary movement."

Employ: See USE.

EMPTY: To discharge contents; to transfer by removing. "Empty tank into overflow tank."

ENERGISE: To supply power. "Energise circuit."

ENGAGE: To cause to interlock or mesh. "Engage threads of turnbuckle with threads of cable terminal."

Note: For circuit breakers, use "close."

ENTER: To go or come in. "Enter compartment through main access."

ERECT: To put up by fitting together. "Erect temporary platform."

ESTABLISH: 1. To set on a firm basis. "Establish safety rules." 2. To perform actions necessary to meet stated conditions. "Establish communications with Control Room."

ESTIMATE: To judge or determine the size, extent, or nature of. "Estimate amount of cleaning solvent which will be necessary."

EVALUATE: To assess; to determine the importance of; to appraise a situation (implies technical knowledge). "Evaluate current plant status."

Examine: See INSPECT.

EXCEED: To go beyond a limit. "Do not exceed a pressure of 400 psi."

EXIT: To go out or away. "Exit building through security doors."

EXTEND: To cause to be drawn out to greater length. "Extend adjustable leg to full length."

EXTINGUISH: To cause to cease burning. "Extinguish any fires using CO2 type extinguisher only."

Extract: See REMOVE.

Fasten: See ATTACH. See SECURE.

FABRICATE: To construct from standardised parts. "Fabricate rig pins from 0.25-inch rod."

Figure: See CALCULATE.

FILE: To rub smooth or cut away with a file (i.e., a tool with cutting ridges for forming or smoothing surfaces). "File one end of rod to a point."

FILL: To put into as much as can be held or conveniently contained or to a specified level; to flood; to replenish. "Fill tank with pure water."

Find: See LOCATE. See DETERMINE.

Flood: See FILL.

FLUSH: To pour liquid over or through; to wash out with a rush of liquid. "Drain and flush hydraulic system."

FOLD: To lay one part over another part; to reduce the length or bulk by doubling over. "Fold sides of curtain on creases."

FOLLOW: To comply; to accept as authority; to obey; to conform to directions or rules. "Follow directions specified in Fig. 1."

FORCE: To exert strength or power to overcome resistance. "Force pin into slot as far as possible."

Appendix B: Example Constrained Word List and Constrained ...   65

FORM: To give a particular shape to; to shape or mould into a certain state; to make up. "Form the compound so that it will fill the hole completely."

Furnish: See PROVIDE.

GAG: To install restraining devices (on relief valves) which prevent operation. "Gag relief valves within test boundary."

GIVE: To put into the possession of another. "Give keys to shift supervisor."

GO TO: 1. To proceed to; to transport oneself to a given destination. "Go to local panel and position switches appropriately." 2. To discontinue use of present procedure and execute another procedure. "Go to MP-1234, Startup Procedure."

Note: Use "refer to" when the user will return to the originating procedure.

GRIND: To pulverise, polish, wear down, sharpen, or smooth by use of a machine or device. "Grind weld to a smooth and even finish."

GROUND: To connect a current, wire, or a piece of electrical equipment to a land or other specified surface. "Ground servicing cart."

GUIDE: To manage or direct the movement of, especially under conditions of close tolerances. "Carefully guide wedge through valve body opening."

HANDLE: To manipulate (load, turn, raise, etc.) objects and equipment manually or with specially designated equipment, such as hoists. "Handle valve stem carefully."

HANG: To fasten to some elevated point without support from below; to suspend. "Hang wiring from temporary overhead hooks."

HEAT: To cause to increase in temperature. "Heat solvent before use."

Help: See ASSIST.

IDENTIFY: 1. To establish the identity of. "Identify components by name and function." 2. To classify a supply item; to note. "Identify component to be ordered from supply."

IF: Establishes a prerequisite which must be met before performing a step. Identifies conditions for operational actions.

Illuminate: See LIGHT.

IMMERSE: To plunge into something that surrounds or covers, especially to plunge or dip into a fluid. "Immerse component in solvent."

IMPLEMENT: To commence a required program or series of procedures. "Implement Emergency Plan."

IMPROVE: To make greater in amount or degree; to make better. "Improve procedures whenever feasible."

Increase: See RAISE. Do not use "increase" because of oral communication problems.

INFLATE: To fill with a given amount of gas or air. "Inflate bladder to desired pressure."

Inform: See NOTIFY.

INITIAL: To affix one's initials. "Initial data sheet when completed."

Initiate: See START.

INSERT: To put or thrust in, into, or through. "Insert wire through hole in panel."

INSPECT: To examine or review present condition. (Method of inspection should be included.) "Visually inspect components for wear, deterioration, or defects."

INSTALL: 1. To perform operations necessary to properly fit an equipment unit into the next larger assembly or system. "Install rocker assembly." 2. To place and attach; to reinstall. "Install nuts on bolts."

Note: For wiring a circuit, use either "install wiring" or "wire." For safety wiring use "install safety wire." For screws, use "install screws" rather than "screw."

INTERCHANGE: To put each in the place of the other. "Interchange printed circuit cards A2 and A3."

ISOLATE: 1. To remove from service. "Isolate letdown." 2. To shut off or separate segments of piping systems. "Isolate feed-water system by shutting valves X and Y."

Join: See CONNECT.

Keep: See MAINTAIN. See AVOID.

LATCH: To catch with a device which holds a door when closed, even if not bolted. "Close and latch containment doors."

Let: See ALLOW.

LIFT: To exert effort to overcome resistance of weight. "Lift test pump to position on platform."

LIGHT: To cause to illuminate. "Light test area using temporary lights."

LINE UP: To bring to proper condition for use. "Line up system for Train A."

LISTEN: To pay attention to sound. "Listen to the pump while it is operating."

LOAD: To place in or on a means of conveyance; to place supplies or components on a vehicle. "Load and secure components on specified truck."

LOCALLY: Refers to actions taken in remote plant areas. "Locally open header to pump isolation valves."

LOCATE: To find, determine, or indicate the place, site, or limits of. "Locate No. 9 fitting."

LOCK: 1. To hold fast or inactive; to fix. "Lock throttle after it has been properly set." 2. To fasten the lock of. "Lock electrical cabinet."

LOCKOUT: To place a control switch or device in a position/condition to be out of service. "Lockout switches A and B."

LOOSEN: To release from restraint; to cause to become less tight fitting. "Loosen lock nut on relief valve."

LOWER: To cause to move down; an action to decrease a parameter. "Lower stem into valve body."

LUBRICATE: To put lubricant on specified locations. "Lubricate pump bearings."

MAINTAIN: 1. To hold or keep in any particular state or condition. "Maintain record of lost supplies." 2.

To take appropriate actions to prevent fluctuation or change. "Maintain test procedure for 15 min."

MANUALLY INITIATE: Control room operator action to activate a function when plant conditions prevent normal automatic initiation.

MANUALLY TRIP: Control room operator action to activate a Reactor Trip or stop an operating piece of equipment such as a pump.

MARK: To label; to provide with an identifying or indicating symbol. "Mark each component before removing it." Note: If marking is to be done on a tag, use "tag."

MATCH MARK: To mark the relative positions of two or more components. "Match mark bonnet assembly to valve body."

Mate: See CONNECT.

MAY: Indicates acceptable or suggested methods. Denotes permission (neither a requirement nor a recommendation).

MEASURE: To determine the dimensions, capacity, or amount by use of standard instruments or utensils. "Measure voltage drop across each unit of resistance."

MIX: To combine or blend into one mass. "Mix resin slurry."

MODULATE: To adjust a valve using a controller to establish a required parameter. "Modulate steam bypass valve controller to commence cool-down."

MONITOR: To continually or periodically attend to displays to determine equipment condition or operating status; to observe current trend. "Monitor indicator for change in pressure."

MOUNT: To attach to a support or specified location. "Mount pressure gauge in its housing."

MOVE: To change the location or position of. "Move valve to a clean area for disassembly."

NEUTRALISE: To destroy the effectiveness of; to nullify; to make chemically neutral or electrically inert. "Neutralise solution before applying to equipment surface."

Note: *See* OBSERVE. *See* IDENTIFY.

NOTIFY: To make known to; to give notice or report the occurrence of; to inform specified personnel; to advise; to communicate; to contact; to relay. "Notify shift supervisor before performing test."

NUMBER: To affix numbers on the pages of the procedure or enclosures. "Number pages of enclosure."

OBSERVE: 1. To conform one's actions or practice to. "Observe precautions." 2. To watch or monitor; to visually take note of; to pay attention to. "Observe indicator when pressure reaches test pressure."

OBTAIN: To gain or attain. "Obtain necessary supplies before starting on maintenance."

OPEN: 1. To move from closed position; to move to the unobstructed position by turning in an appropriate direction to permit access or flow. "Open relief valve." 2. To break an electrical connection which removes a power supply from an electrical device. "Open circuit breakers." 3. To make available for entry or passage by turning back, removing or clearing away. "Open top maintenance access."

OPERATE: 1. To control equipment in order to accomplish a specific purpose. "Operate fire extinguisher." 2. To open and close valves as necessary to perform the intended function. "Operate vent valve to release non-condensable gases." 3. To place pumps or breakers in the state necessary for them to perform their intended function. "Operate standby pump for three minutes."

OR: Establishes that each of a series of actions is equally preferable. Indicates alternatives.

Order: See ARRANGE.

Organise: See ARRANGE.

Orient: See POSITION.

OVERHAUL: The act of disassembling equipment units down to all removable parts; cleaning; critically inspecting, repairing, restoring and replacing where necessary; assembling, adjusting, aligning, recalibrating and verifying operational readiness by test or checkout. "Overhaul No. 2 pump."

PACK: To fill completely with grease. "Pack bearings."

PAINT: To apply colour or pigment (suspended in suitable liquid) to a surface. "Paint all exposed surfaces."

PATCH: To mend, cover, or fill up a hole or weak spot. "Patch tubes where necessary."

PER: 1. By means of. "Per our agreement." 2. To, for, or by each. One quart of cleaner per gallon of water.

PERFORM: To do, carry out or bring about; to accomplish; to effect; to reach an objective. "Perform hydrostatic test of steam generator."

PLACE: To put or set in a desired location or position; to locate. "Place test equipment next to electrical cabinet but away from traffic areas."

PLUG: To provide with a device for closing off the end of a tube which has a female fitting; to insert or install plugs. "Plug all lines which have exposed female fittings."

POSITION: To put or set in a specific configuration, place or orientation; to locate; to reset. "Position test equipment so that it can be seen by both technicians."

POUR: To cause to flow in a stream. "Pour drainage into a waste reservoir."

PREPARE: 1. To make ready; to arrange things in readiness; to set up. "Prepare surface for paint." 2. To put together or make ready for a maintenance activity. "Prepare additional data sheets as necessary."

PRESS: To act upon with steady force. "Press blower start button." Note: For circuit breakers, use "CLOSE."

PRESSURISE: To apply pressure within by filling with gas or liquid; to re-pressurise. "Pressurise first stage chamber."

PREVENT: To keep from happening or existing. "Prevent oil from spilling over on components."

PROVIDE: To furnish; to supply what is needed; to equip. "Provide a flashlight for man B."

Appendix B: Example Constrained Word List and Constrained ...

PULL: To exert force upon an object so as to cause motion toward the force. "Pull out knob No. 6 on oxygen servicing cart."

Note: For circuit breakers, use "open."

PUMP: 1. To raise or lower by operating a device which raises, transfers, or compresses fluids by suction, pressure, or both. "Pump out overflow from catch pan." 2. To move up and down or in and out as if with a pump handle. "Pump engine primer knob."

PUNCTURE: To pierce with a pointed instrument or object. "Be careful not to puncture tube walls while inserting instrument."

PURGE: To free of sediment or trapped air by flushing or bleeding. "Purge fuel lines."

PUSH: To move away or ahead by steady pressure. "Push access door until it latches in position."

PUT: To deposit or leave. "Put tools out on the bench."

Note: Use "store" instead of "put away" for depositing or leaving in a specified place for future use.

RACK IN: To place an electrical breaker in service by physically connecting it to its associated power source. "Rack in breaker A to 420-V bus."

RACK OUT: To remove an electrical breaker from service by physically disconnecting it from its associated power source. "Rack out breaker A from 420-V bus."

RAISE: 1. To move or cause to be moved from a lower to a higher position; to elevate. "Raise control level to RELEASE position." 2. To make greater in size, amount or intensity. "Raise tank pressure to 400 psi."

READ: To interpret the meaning of by visual observation. "Read ammeter."

Readjust: See ADJUST.

Ready: See PREPARE.

Reassemble: See ASSEMBLE.

RECALL: To call back. "Recall parts which have not been modified."

Recap: See CAP.

Recapitulate: See REPEAT.

RECEIVE: To come into possession of; to get. "Receive supplies as they arrive."

RECOGNISE: To perceive to be something previously known or diagnosed. "Recognise troubles through evaluation of engine operational checks."

RECOMMEND: To urge the acceptance or use of. "Recommend procedure changes where appropriate."

Reconnect: See CONNECT.

RECORD: To document a specified condition or characteristic. "Record discharge pressure on data sheet."

REDUCE: To cause to be diminished in strength, density or value; to decrease. "Reduce pump flow."

REFER: 1. To call or direct attention to a supplement. "Refer to Fig. 1 for part numbers." 2. To give direction to perform actions in another procedure and return to the originating procedure. "Refer to MP-1234, Startup Procedure, steps 7.1.1 through 7.3.12."

Note: Use "go to" when the user will not return to the originating procedure.

Regulate: See CONTROL.

Reinflate: See INFLATE.

REJECT: To refuse to have, use or take for some purpose. "Reject components which show excessive wear."

Relay: See NOTIFY.

Reinstall: See INSTALL.

RELEASE: 1. To set free from an inactive or fixed position; to unlock. "Release AUTO hold bar to provide manual control." 2. To let go of. "Release tensiometer handle." 3. To set free from restraint or confinement. "Release pressure."

REMOVE: 1. To perform operations necessary to take an equipment unit out of the next larger assembly or system. "Remove bleed air shutoff valves." 2. To take off or eliminate. "Remove paint." 3. To take or move away; to extract. "Remove covers." 4. To take off devices for closing off the end of a tube; to uncap or unplug; to break. "Remove caps (plugs) from all hydraulic lines." Note: For screws, use "remove" rather than "unscrew."

REPAIR: To restore equipment to operable condition by a means other than total replacement of a part. "Repair connector by re-soldering leads." Note: Repair includes such methods as gluing, reattaching, patching, welding, splinting, building up (a surface), sanding smooth, straightening, re-soldering. Repair does not involve isolation of a fault. In accomplishing repair, no items are drawn from supply except consumables, attaching parts or reinforcing parts.

REPEAT: To make, do or perform again; to recapitulate. "If keys do not engage lugs, remove assembly and repeat procedure."

REPLACE: To change or substitute serviceable equipment for malfunctioning, worn out or damaged equipment. "Replace switch contact points."

Replenish: See FILL.

REPORT: To describe as being in a specific state. "Report conditions of worn or frayed wires to I&E Maintenance."

Repressurise: See PRESSURISE.

REQUEST: To ask for. "Request further information if necessary."

Reset: See SET. See ADJUST. See POSITION.

RESTORE: To bring back or put back into a former or original state. "Restore hydraulic pressure."

RETRACT: To draw back or in. "Retract locking pins by turning lock screw (2) COUNTERCLOCKWISE."

RETURN: To bring, send or put back to a former or proper place. "After testing, return throttle valves to preset positions."

REVIEW: To examine again, to go over or examine critically or deliberately. "Review test data and signature sheet to verify all blanks have been filled in."

RINSE: To clean (as from soap used in washing) by clear water. "Rinse battery after cleaning it with soda water solution."

RISE: Describes an increase in a parameter as the result of an operator or automatic action. "Observe rise in condensate tank."

ROPE OFF: To partition, separate or divide by a rope. "Clear and rope off an area around the generator and post warning signs."

ROTATE: 1. To cause to revolve about an axis or centre; to turn. "Rotate traveling screens 90 degrees." 2. To hand-rotate a pump before energising. "Rotate pump."

SCAN: 1. To examine closely. "Scan this document for any mistakes."

SCRAPE: To remove an outer layer of an object by forceful strokes of a rough or sharp instrument. "Scrape the sticker paper off the glass."

SCREW: To turn or twist. "Screw it into the hole." Note: For screws, use "install screws" rather than "screw."

SCRUB: To rub hard in order to clean or remove a substance. "Scrub the oil out of the treads."

SEAL: Joining two items in a way that prevents leakage, or an airtight closure. "Seal the envelope."

SECURE: To make firm or tighten. "Secure the ladder to prevent movement, while climbing."

SELECT: To take as a choice among several options. "Select the correct file."

SEPARATE: 1. To keep apart. "Separate the oil and water-based paints." 2. To come apart. "Tighten the bolt or the component will separate.

SET: To put in a specified position. "Set the can on the deck."

SHAKE: To move an object in rapid motion up and down or left and right. "Shake the can of paint before opening."

SHALL: To be able to do something. "I shall walk."

SHIFT: 1. To change gears. "Shift into second gear." 2. To change position. "Shift the can a little to the right." 3. A period of time at work. "He is the shift supervisor."

SHOW: To be or become visible. "Look at the numbers shown on the screen."

SHOULD: Used to express obligation or probability. "He should call by one o'clock."

SHUT DOWN: To cease to operate. "Shut down your computer when you are done using it."

SIGN: To write one's signature. "Sign your name to the document."

SIGNAL: An indicator. "The alarm signals danger."

SLIDE: To move over a surface with smooth continuous contact. "Slide the can across the deck."

SLIP: To move smoothly and easily. "Slip the nut into place."

SPARE: To use with restraint. "Spare the oil."

SPECIFY: To state in detail. "Specify which route to use."

SPILL: To run or fall out of a container. "The paint spilled out of the can."

SPIN: To rotate rapidly. "Spin the hose around."

SPRAY: 1. Water or other liquid moving in a mass of dispersed droplets. "The ocean spray may cause the deck to be wet at times." 2. To disperse a liquid in a mass of droplets. "Spray the water on the deck."

SQUIRT: To issue forward in a thin jet. "Squirt the water on the ladder."

STANDBY: An item kept in reserve for use when needed. "Switch to the standby pump."

START: To begin an activity or movement. "Start operating the pumps."

STIR: To disturb relative position of particles or parts of, especially by a continued circular motion. "Stir sample before performing conductivity test."

STOP: To perform actions necessary to cause equipment to cease or suspend operation. "Stop PUMP."

STORE: To deposit or leave in a specified place for future use; to stow; to put away. "Store equipment covers after maintenance activity is completed."

Stow: See STORE.

SUBMIT: To make available; to offer; to deliver. "Submit completed Work Request to Shift Supervisor."

SUPPORT: To hold up or provide a foundation or props for. "Support assembly at both ends."

SURVEY: To examine comprehensively. "Survey entire equipment surface."

SUSPEND: To stop action; to leave system as it stands. "If pressure rises above 400 psi, suspend activity until pressure drops below 380 psi."

SYNCHRONISE: To establish phase-to-phase alignment. "Synchronise Bus A with Bus C."

Tabulate: See RECORD.

TAG: To provide with an identifying or indicating symbol with or as if with a tag (i.e., a cardboard, plastic or metal marker used for identification or classification); to label; to attach or connect a tag to; to mark.

"Tag each hydraulic line before removing it."

TAKE: 1. To get into or carry in one's hands or one's possession; to deliver. "Take valve to a clean area for disassembly." 2. To get or find out by observation or special procedures; to obtain. "Take a reading on outside circle of tensiometer."

TAP: To strike lightly. "Tap eye of cotter pin to seal it."

TEST: To perform specified operations to verify operational readiness of a component, subcomponent, system or subsystem; to check out. "Test accuracy of indicator as follows:"

THEN: Indicates actions to be performed after stated conditions have been established.

THROTTLE: To operate a valve in an intermediate position to obtain a certain flow rate. "Throttle valve for three minutes."

TIE: To fasten, attach, or close by means of a line or cord. "Tie support ropes to equipment."

TIGHTEN: To perform necessary operations to fix more firmly in place. "Tighten all screws." Note: To tighten to a specific torque value, use "torque" nuts.

TORQUE: To apply a specified amount of force to produce a rotation or twisting motion to fix more firmly in place; to tighten. "Torque nut to 100 in-lbs." Note: Torque (noun): length of spanner handle times applied force.

TRACE: To follow or study out in detail or step by step. "Trace wiring from breaker to faulty component."

TRANSFER: To convey, transport, transmit, or cause to pass from one place to another. "Transfer radioactive test source to test area."

TRIM: To free of excess or extraneous matter by cutting. "Trim leads."

TRIP: To manually activate a semi-automatic feature. "Trip breaker."

TROUBLESHOOT: To localise and isolate the source of a malfunction or breakdown. "Troubleshoot pump control circuit."

Turn: See ROTATE.

Uncap: See REMOVE (cap).

UNLOCK: To unfasten the lock of; to open. "Unlock electrical cabinet."

Unplug: See DISCONNECT. See REMOVE (plug).

UNSCREW: To loosen or withdraw by turning in the proper direction. "Unscrew adapter." Note: For screws, use "remove screws."

UNWIND: To cause to uncoil or unroll. "Unwind hoses from hose rack."

USE: To put into action or service; to avail oneself of; to carry out a purpose or action by means of; to utilise; to employ. "Use only antimagnetic fasteners."

Utilise: See USE.

VENT: To permit gas or liquid under pressure to escape at a vent. "Vent non-condensable gases using RCS vent valves."

VERIFY: 1. To make sure by taking necessary or appropriate actions. "Verify discharge pressure is stable." 2. To establish the truth or accuracy of. "Verify readings before recording them."

Note: *Use of* "verify" *requires a step signoff.*

WAIT: To suspend activity in a sequence of activities until a given condition occurs or a given time has elapsed. "Wait five minutes before performing next task."

WASH: To cleanse by or as if by the action of liquid; to remove (dirt) by rubbing or drenching with liquid. "Wash battery with cleaning solution and a stiff brush."

Watch: See OBSERVE.

WEIGH: To measure the heaviness of something with use of a scale. "Weigh material and record weight in Attachment 1."

WHEN: Indicates certain condition must be established before the step can be performed.

WILL: Indicates requirement.

WIPE: To rub with something soft for cleaning or drying. "Wipe dry."

WIRE: To provide with wire; to use wire on; to install wiring. "Wire circuit."

WITHDRAW: To take back, away or out. "Withdraw bar magnet from centre of coil."

WRAP: To wind or coil as to encircle or cover something. "Wrap wire around terminal."

Appendix C: Analysis Technique

This appendix outlines optional analysis techniques that can be carried out to aid in the development of instructional materials. These analysis techniques, the amount of preliminary work (e.g., needs assessment, discovery, etc.), and the number and type of people involved (e.g., SME's) all depend on the complexity of the system and the system's role in overall vessel/offshore installation environmental and personnel safety.

Conducting a Needs Assessment

After all relevant information has been gathered, it may be necessary to perform a needs assessment. Some issues that prompt a needs assessment are:

- Concerns for safety (near miss / audit findings)
- Complexity of the task(s); and
- Regulatory requirements.

The needs assessment is performed to determine the information needs of the user. The output is the basis for the document's development. The scope, impact, and complexity of work activities to be covered should be considered, especially in cases where materials are to be developed for new companies. However, it is not always necessary to perform a needs assessment, such as in circumstances where existing documentation and experience can be used in place of a formal needs assessment. When there are serious safety or environmental impact concerns, a needs assessment is usually justified. The purpose of a needs assessment is to identify and describe the elements of a situation or task that may require development of new materials or revision to an existing document. This assessment also helps identify the activities and performance expectations of tasks related to the procedure. This is done by assessing the equipment involved and the process to be completed, measuring existing performance against the accepted or desired principles (called the gap analysis) and identifying the causes of the gap between actual performance and the desired performance level. This information is accumulated during discovery.

Discovery

Discovery is the process for accumulating the information needed to conduct a Gap Analysis as it relates to operating equipment or providing a service. To identify the gap, the associated equipment and existing personnel competencies are assessed. In the case of new documents requiring new skills, the gap may be wide. To facilitate discovery, the following should be done:

- Equipment requiring manipulation by personnel should be assessed
- Tasks requiring performance should be identified
- People skill sets and abilities as well as equipment condition and availability should be evaluated
- Metrics should be developed to quantify the performance levels that contribute to the success of the organisation
- Existing instructional materials should be collected and analysed; and
- Interviews should be conducted with all stakeholders, including the persons who will use the documents.

In addition to measuring existing competencies and assessing the related equipment, other areas of concern during discovery are:

- Identifying internal and external constraints such as, regulatory, monetary, or any other restriction or limitation imposed on the document need
- Assessing all operating modes such as daily operation, startup, and shutdown
- Using essential tools such as questionnaires, interviews, and direct observation of the process
- Collecting data in a coherent and organised manner so that it can be effectively analysed; and
- Developing a roadmap that identifies the areas where performance does not meet the stated goals or acceptable standards of the organisation (primarily refers to existing materials requiring revision).

Identifying and Prioritising Opportunities

To provide direction for the document development process, the information collected during discovery should be organised and prioritised. It may be helpful to divide the information gathered into:

(1) Opportunities; and
(2) Constraints.

Associating Opportunities with Constraints

Constraints to the process are conditions that exist within the process but cannot be changed. Government regulation, environmental stewardship, and cost are all examples of constraints. Opportunities such as enhancing people's competencies through training, should be assessed in light of the related constraints, which may include a limited budget for training.

Priority Analysis

Data gathered during discovery should be prioritised for analysis. The needs identified by the needs assessment should be directly supported by the prioritised data.

Developing a Useable List of Priorities

A document should be generated listing all issues and opportunities ranked in priority order. Proposed solutions may be included, but the primary concern is to develop a useable list of priorities that can be analysed and addressed by the needs assessment team.

Action Plan/Recommendations

The information gathered during discovery should be compiled into a set of recommendations or necessary actions that directly address the need for instructional materials and are supported by the data collected.

Available Resources

The action plan should be achievable within the bounds of available resources, or it should specify additional resources (personnel, equipment, capital, etc.) that are necessary to be implemented. As an example, the planned replacement of equipment nearing the end of its life cycle with different equipment may have prompted a needs assessment, which management considered when budgeting for a given time-frame. The needs assessment should determine the feasibility of realising the plan and developing the required materials using budgeted resources.

Final Product

The final product of the needs assessment should be a document that lists the identified training and document needs, along with any possible solutions that were proposed.

Conducting a Task Analysis

A task analysis focuses on understanding the functions and tasks that personnel will perform with the system in terms of:

(1) The inputs and outputs for those tasks with regard to required information and control actions
(2) Feedback provided to an operator after the manipulation of a control
(3) Dependencies between tasks
(4) Sequence of task performance
(5) Time and frequency of performance
(6) Accuracy requirements
(7) Criticality or importance of the tasks; and
(8) Error tolerance/risk/criticality.

By performing a task analysis, the document developer can get an understanding of how a workstation or work area is designed and arranged to support reliable task performance. Task analysis also focuses on understanding the flow and sequence of human activity as work is performed. It can be used to understand how personnel move between controls, displays, consoles, panels, workstations, equipment, and work areas in performing a series of tasks. This information is used to develop instructional materials to promote accuracy and efficiency when carrying out the tasks. Figure C.1, "Example Form for Task Analysis" provides an example table that can be used when conducting a task analysis for instructional documents. The table can be tailored to fit the needs of individual organisations. However, the first three columns (i.e., "Step #" "Description" and "Activity") should be used regardless, as they provide the more vital information with regard to document development.

Conducting a Risk Analysis

A risk analysis involves identification, evaluation, and estimation of the levels of risks involved in a vessel or offshore structure, the comparison against benchmarks or standards, and the determination of an acceptable level of risk. A risk analysis is the process used to develop a risk assessment. The risk analysis and risk assessment are steps in a Risk Management Program. A general discussion of risk analysis is presented here. The risks

Appendix C: Analysis Technique 79

Step #	Description	Activity	Information Required	Technical Limits	Hazards	Complexity	Criticality	Controls & Displays	Job Aids and Tools	Time or Sequence	Notes

Fig. C.1 Example form for task analysis for instructional materials

associated with equipment and processes that will be described in instructional materials should have already been identified in a task analysis.

Conducting a HAZOP Study

For offshore installations such as drilling rigs and operating platforms, a Hazard and Operability (HAZOP) study may be used, and concepts applied to shipboard facilities as well. A HAZOP study is a type of risk assessment performed during the design stage of a new vessel or offshore installation or addition to an existing vessel or offshore structure, although HAZOP studies can be used at various stages of a vessel or offshore structure's life. It is briefly mentioned here to call attention to document writers so that identified hazardous operations are clearly described in their associated task steps.

Job Hazards Analysis

Job hazards analysis (JHA) is a process used to review the hazards associated with a task and reduce the likelihood of harm to personnel involved in a job task. A JHA is important to help verify that adequate safeguards are in place to control hazards. During a JHA analysis, the individuals tasked with performing a job identify the basic job steps and tasks. For every step, associated hazards are reviewed. For every hazard identified, there is confirmation of an effective control or safeguarding mechanism in place to control the hazard. Example controls could be safe operating procedures, PPE, hardware, alarms, etc.

Summary of Pre-procedure Development Activities

The initial information collected by the SMEs who are brought together for the purpose of developing instructional materials includes source documentation such as regulatory documents, related engineering drawings, and vendor documentation. After this is accomplished, the SMEs determine if the extent of their work necessitates a needs assessment, which would allow them to assess the needs from a general point of view. This can often be accomplished without a formal needs assessment. After the SMEs understand their general needs, they then decide if a task analysis is necessary to isolate the individual tasks involved in the larger process. In some cases, the SMEs could elect to put the associated equipment and processes through a risk assessment to focus on the potential hazards associated with performance of the various tasks that have been identified. The output of these processes should be a Document Requirement List that may include the subject and scope of the proposed materials. After all pre-development activities are completed, the tasks have been identified, and the subject of the materials have been established and confirmed, document development can begin as described in Chap. 3, "Writing Instructional Documents".

Appendix D: Examples of Instructional Materials

This appendix contains "real" examples that are currently in use on vessels and/or offshore installations. The examples are presented in their entirety with only proper names, etc., removed. The examples presented herein are intended solely to assist the reader in the methodologies and/or techniques discussed and are for illustration and general information purposes only. They are not to be relied upon or intended to be a substitute for the reader's particular decision-making requirements or serve as professional advice. This guidance does not and cannot replace the analysis and/or advice of a qualified professional. It is the responsibility of the reader to perform their own assessment and obtain professional advice. The subject matter covered by the examples is considered to be pertinent at the time of publication, but it is still possible that all or part of the information contained herein may be invalidated as a result of subsequent legislation, standards, methods, and more updated information. The reader assumes full responsibility for compliance with any rules, regulations, industry standards, and other laws applicable in the reader's jurisdiction.

Inappropriate Procedure

The following procedure is an example of a literal translation to English from another language. As this procedure has been translated, it would be exceedingly difficult for a person to carry out the steps appropriately. Rating of starting compensating transformer is short time. It is necessary to pay attention to the following subjects:

(1) Number of consecutive starting time are to be within 4 times; and
(2) After 4 times starting in succession 240 min of pause is necessary.

Better Procedure

A better way to present this procedure could be the following:

Only attempt to start the Compensating Transformer four (4) times.

If Compensating Transformer has not started after four (4) attempts, you must wait four (4) hours before attempting again.

Inappropriate Procedure

The following procedure is a sample incinerator operating procedure. The overall format could make it difficult for users to follow. The font is small, steps are not clearly defined (i.e., would the user know to do the steps in any particular order), and the Danger is located at the end of the procedure with no emphasis.

- The incinerator operator shall be professionally trained
- Wear fire retardant apron, elbow gloves and face shield while loading or operating the incinerator
- Request permission from the bridge to commence incineration
- Check the eye glass to verify there is no fire present before opening door
- Verify the incinerator is ignited with the automatic starter
- Verify the door is closed and latched after rubbish is placed in the incinerator and that the load does not exceed 1/2 full level
- Verify the door is not opened during incineration
- Verify the contents are left to burn for at least one day
- Verify the furnace is cleaned out when it is 1/4 full of ash and that ashes are not removed until they have cooled off
- DANGER: Disposal of aerosols and batteries in the incinerator is prohibited. Dispose only in designated receptacles.

Better Procedure

The Danger should be the first thing that the user reads, and the steps should be numbered sequentially so that users understand the logical order of all steps in the task. A better way to present the procedure could be as follows.

Incinerator Operating Procedure

DANGER: Disposal of aerosols and batteries in the incinerator is prohibited. Dispose only in designated receptacles.

Description.

(1) Wear fire retardant apron, elbow gloves and face shield while loading or operating the incinerator.
(2) Request permission from the bridge to commence incineration.

(3) Before opening the door, check the eye glass to verify there is no fire present.
(4) Verify the incinerator is ignited with the automatic starter.
(5) Verify the door is closed and latched after rubbish is placed in the incinerator and that the load does not exceed 1/2 full level.
(6) Verify the door is not opened during incineration.
(7) Verify the contents are left to burn for at least one day.
(8) Verify the furnace is cleaned out when it is 1/4 full of ash and that ashes are not removed until they have cooled off.

Example Checklist

Below is an example checklist. Checklists are a type of job aid that can be used for complex isolations (e.g., watch turn-over) to verify that the actions have been performed and/or the events have occurred.

Pilotage

- Immediately on arrival on the bridge, has the pilot been informed of the ship's heading, speed, engine setting, and draft?
- Has the pilot been informed of the location of lifesaving appliances provided onboard for his use?
- Have Details of the proposed passage plan been discussed with the pilot and agreed with the Master including:
- Radio communications and reporting requirements
- Bridge watch and crew stand-by arrangements
- Deployment and use of tugs
- Berthing/anchoring arrangement
- Expected traffic during transit
- Pilot changeover arrangements, if any
- Fender requirements
- Has a completed pilot card been handed to the pilot and has the pilot been referred to the Wheelhouse Poster?
- Have the responsibilities within the bridge team for the pilotage been defined and are they clearly understood?
- Has the language to be used on the bridge between the ship, the pilot, and the shore been agreed?
- Are the progress of the ship and the execution of orders being monitored by the Master and the officer of the watch?
- Are the engine room and the ship's crew being regularly briefed on the progress of the ship during pilotage?
- Are the correct lights, flags, and shapes being displayed?

Bibliography

Campbell, J. and Zimmerman, C. (1988). Fundamentals of Procedure Writing (2nd ed.) Maryland: GP publishing, Inc.
Centre for Chemical Process Safety. Guidelines for Risk Based Process Safety. New Jersey: Author.
Convention on the International Regulations for Preventing Collisions at Sea, 1972 (COLREGS), http://www.imo.org/Conventions/mainframe.asp?topic_id=257&doc_id=649.
Health and Safety Executive. (1997). Improving Compliance with Safety Procedures, Reducing Industrial Violations. Sudbury, Suffolk, England: HSE Books.
International Chamber of Shipping. (2007). Bridge Procedures Guide (4th ed). London: Author
International Maritime Organisation. (1972).
Technical Writing Guidelines, The Natchez Group Inc., dba TechProse., 1 September 2004.
Technical Writing Made Easier, Bernhard Spuida, March 2002.

SPRINGER NATURE

GPSR Compliance

The European Union's (EU) General Product Safety Regulation (GPSR) is a set of rules that requires consumer products to be safe and our obligations to ensure this.

If you have any concerns about our products, you can contact us on ProductSafety@springernature.com

In case Publisher is established outside the EU, the EU authorized representative is:

Springer Nature Customer Service Center GmbH
Europaplatz 3
69115 Heidelberg, Germany

The manufacturer's authorised representative in the EU is Springer Nature Customer Service Centre GmbH, Europaplatz 3, 69115 Heidelberg, Germany. If you have any concerns regarding our products, please contact ProductSafety@springernature.com

Printed and bound by CPI Group (UK) Ltd, Croydon, CR0 4YY
26/03/2026
02078987-0003